Quality and Production of Forage

Quality and Production of Forage

Special Issue Editor
Vincenzo Tufarelli

MDPI • Basel • Beijing • Wuhan • Barcelona • Belgrade

MDPI

Special Issue Editor
Vincenzo Tufarelli
University of Bari Aldo Moro
Italy

Editorial Office
MDPI
St. Alban-Anlage 66
4052 Basel, Switzerland

This is a reprint of articles from the Special Issue published online in the open access journal *Agriculture* (ISSN 2077-0472) from 2018 to 2019 (available at: https://www.mdpi.com/journal/agriculture/special_issues/forage).

For citation purposes, cite each article independently as indicated on the article page online and as indicated below:

LastName, A.A.; LastName, B.B.; LastName, C.C. Article Title. *Journal Name* **Year**, *Article Number*, Page Range.

ISBN 978-3-03921-674-1 (Pbk)
ISBN 978-3-03921-675-8 (PDF)

Cover image courtesy of Eugenio Cazzato.

Contents

About the Special Issue Editor

Vincenzo Tufarelli is a Researcher in Animal Nutrition at the Department of Emergency and Organ Transplants (DETO), Section of Veterinary Science and Animal Production of the University of Bari Aldo Moro, Italy. He has considerable experience in animal and poultry science, with particular interest in nutrition and feed technology. He is involved in many research collaborations, including with international institutions, in the field of animal science and feed quality. He is serving as an Editorial Board Member and peer reviewer for many indexed journals, and he is the author of more than 150 scientific papers published in international journals and the proceedings of national and international conferences. He received the Italian Professorship Qualification as Associate Professor for the scientific sector 07/G1 (Animal Science and Technology).

Preface to "Quality and Production of Forage"

Forage quality is defined in several ways, but is often inadequately understood. It is a simple concept, yet encompasses much complexity. In recent years, advances in plant and animal breeding, the introduction of new products, and the development of new management approaches have made it possible to increase animal performance. However, for this to be realized, there must be additional focus on forage quality. Producing suitable-quality forage requires knowing the factors affecting forage quality, then exercising management accordingly. This book presents cross-discipline studies covering many aspects ranging from forage production and nutrition to animal feeding with the aim of disseminating information suitable to improve forage quality and utilization. Moreover, the purpose of this book is also to provide information about forage quality and forage testing that can be used to increase animal performance and producer profits.

Vincenzo Tufarelli
Special Issue Editor

agriculture

MDPI

Review

Feeding Forage in Poultry: A Promising Alternative for the Future of Production Systems

Vincenzo Tufarelli [1,*], Marco Ragni [2] and Vito Laudadio [2]

[1] Department of DETO, Section of Veterinary Science and Animal Production, University of Bari 'Aldo Moro', Valenzano, Bari 70010, Italy

[2] Department of Agricultural and Environmental Science, University of Bari 'Aldo Moro', Bari 70125, Italy; marco.ragni@uniba.it (M.R.); vito.laudadio@uniba.it (V.L.)

* Correspondence: vincenzo.tufarelli@uniba.it; Tel.: +39-080-544-3916

Received: 25 May 2018; Accepted: 8 June 2018; Published: 9 June 2018

Abstract: The present review discusses the existing research findings on the nutritional impact of forages in poultry diet and the significance of forages in sustainable poultry production systems. The nutritional composition and antinutritional factors of the main forages and the pros and cons of feeding forage on poultry meat and egg quality under free-range and organic production systems are also discussed. This review highlights the importance of forages and forage meals in poultry ration, considering that these feedstuffs may have greater value to the success of local poultry production in many regions of the world due to their potential of production.

Keywords: forage; production system; poultry; sustainability; livestock

---

1. Introduction

Organic and free-range poultry have been increasingly available to the average consumer due to increased consumers' demand for meat and egg products [1]. In fact, with the increased intensity of modern production agriculture, many consumers share views that organic and free-range products are in some way better for them, or at least follow production practices that are more conducive to a cleaner, more balanced environment [2]. Organic farming is defined as the approach to agriculture in which the aim is to create environmentally and economically sustainable agricultural production systems [3].

In organic poultry farming, feeding covers an important role, since the dietary requirements of poultry are very different from those of ruminant livestock [4]. Poultry are particularly sensitive to dietary quality because they grow quickly and make relatively little use of bulky fibrous feeds such as pasture or hay [5]. In addition, poultry have specific requirements for essential amino acids, in particular lysine and methionine. In conventional systems, feeds are supplemented with synthetic amino acids, but these are not permitted in organic systems, so alternative sources, such as natural essential compounds from plants and vegetables, have to be investigated [5,6]. It is often difficult and expensive to source these products, and they are major contributors to the high cost of feed. In this regard, there is a growing interest in using novel feed resources, and the utilization of forage for poultry diet could be a sustainable and natural alternative for organic producers, especially under smallholder production system conditions.

Forages and forage meals could be valuable alternative sources of protein for commercial poultry where they are easily available and not expensive [7,8]. Moreover, forage utilization as local feed resource is important to facilitate the transition to 100% organic feed supply for organic poultry meat and egg producers [9]. Therefore, the present review discusses the available research findings on the nutritional impact of forages in poultry diet and the significance of forages in sustainable poultry production systems.

2. Forage Nutritional Composition

The demand for animal protein for human nutrition in the developing world is still rising, especially for poultry products and the cost of feed concentrates for livestock is increasing [3]. Therefore, to meet the nutritive requirements of poultry, it is necessary to identify alternative low-cost feed resources.

Forage based animal production plays a crucial role in the affordable supply of nutrient rich foods for humans [10]. Grassland and forage crops are recognized for their contribution to the environment, recreation and efficiency of animal production [11,12]. To maintain sustainability, it is crucial that such farming systems remain profitable and environmentally friendly while producing nutritious foods of high economical value. Thus, it is pertinent to improve the nutritive value of grasses and other forage plants to enhance animal production to obtain quality food, and it is also vital to develop new forages which are efficiently utilized and wasted less by involving efficient animals [13]. A combination of forage legumes, fresh or conserved grasses, crop residues and other feeds could help develop an animal production system which is economically efficient, beneficial and viable [9].

Livestock animals and forages are two significant factors in many animal production systems [13]. Grasses, legumes and cereals can be tested for their roles for sustainable animal production, and searching for new forages could help reduce the effects of animal production on climate change. In fact, forages can be tested further for their environmental benefits and contribution to the sustainable animal production [14]. Therefore, more research on the suitability of animal species and genotypes to utilize available and alternative feed resources more effectively may also help in developing sustainable forage based animal production.

Forages as feed for monogastrics, including poultry, contribute to improve sustainability of animal production within farming systems [15]: high biomass production in environments where other crops cannot compete; no or limited competition with human food requirements; high levels of protein with a desirable amino acids profile, especially lysine, methionine and other sulfur-amino acids, which for monogastrics adequately balances the limitations of cereal proteins (leaf and grain); and additional benefits from the integration of forages in the farming system [15]. The chemical composition and nutritive value of many forages were widely summarized by the French Institut National de la Recherche Agronomique (INRA) [16].

Plants produce a variety of simple to highly complex anti-nutritional substances, many of which have been identified and characterized [17]. The most common major groups are polyphenols, cyanogenic glycosides, alkaloids, saponins, steroids, toxic proteins and amino acids, non-protein amino acids, phytohemagglutinins, triterpenes, and oxalic acid [18], and are either toxic or act as antinutritive factors. Secondary compounds can exert both anti-nutritional and nutritionally beneficial effects upon forage feeding value. Secondary compounds that occur in both temperate and tropical forages [19]. Lectins are sugar-binding glycoproteins, which are classified as toxic (*Phaseolus vulgaris* and *Canavalia ensiformis*), growth inhibitory (*Glycine max*, *Amaranthus cruentus*, *Phaseolus lunatus*, and *Dolichos biflorus*) [20], or essentially non-toxic or beneficial (seeds of *Vigna subterranea*, *Vigna umbellata* and *Vigna unguiculata*) [20]. Condensed tannins are complex heat-stable phenolic compounds and common in many plants, especially shrub legumes such as *Gliricidia sepium*, *Acacia species*, *Leucaena leucocephala* and *Albizia falcataria*. Polyphenols are a major group often related to taste, odor, and color of animal products [21]. Saponins are found in *Brachiaria decumbens*, *B. brizantha* [22], *Amaranthus hypochondriacus*, *Chenopodium quinoa*, *Atriplex hortensis* [23], and *Medicago sativa* [24]. Alkaloids of legumes such as the bitter-tasting quinolizidine in lupins [25] reduce the feed intake, may affect the liver, and paralyze respiration [26].

The presence of secondary compounds can have a profound effect on both the nutritive value and the feeding value of both temperature and tropical forages and these effects can be beneficial in some instances as well as being detrimental in others [19,27]. To completely understand the influence of these compounds on animals, it is essential to develop a knowledge of their chemical structure and reactivity, particularly with proteins [27].

However, forage plants can be successfully processed to enhance palatability, intake and digestibility, and to conserve, detoxify the antinutritional factors, or concentrate nutrients [19,28]. In particular, milling, pelleting and micronizing processes increase the digestibility in poultry of protein and starch and apparent metabolizable energy values of forages [28,29]. In this regard, some antinutritional factors such as tannins are mainly concentrated in the seed coat, so that hulling is a simple method to remove them [30,31]. There have been various attempts to mix different exogenous enzymes into feeds to reduce antinutrients [32].

3. Poultry Production Systems

Free-range and organic poultry production systems have increased in recent years worldwide, and it is widely known that the suitability of the genetic strain to extensive environments largely affects the animal welfare and the meat and egg characteristics [6,33]. In fact, when poultry are reared under extensive system having access to pasture, health, welfare, meat and egg quality parameters result enhanced [34,35].

3.1. Free-Range System

Free-range defines a production system where the chickens can access outdoor runs for most of their lives [36]. These systems are well known and extensively used by small-scale farmers and, normally, slow-growing broilers are used [37]. However, free-range does not mean that the production methods follow certified organic standards as some consumers may perceive. A conventionally-raised chicken could be labeled as free-range if it were allowed to have access to the outdoors. Free-range production typically follows an alternative rearing practice compared to that of conventional production [38].

The U.S. free-range system definition does not include any specification for how long broilers should remain outdoors and under what conditions they should be raised and fed [37,39]. In addition, there is no specification for appropriate genotypes and no minimum age for slaughter. To label broilers as free-range, farmers in the U.S. must demonstrate that birds have had free access to the outdoors for more than half of their lives. Conversely, the European legislation for free-range systems specifies maximum stocking rates for indoors and outdoors, minimum age at slaughter as well as the feed composition for both laying hens and broilers [5].

Improving poultry welfare is one of the key reasons for the re-development of free-range poultry production in Europe. The European legislation also differentiates free-range into three main systems: free-range, traditional free-range and total freedom (Council Regulation EC No 1234/2007) [40]. In free-range systems, poultry houses are normally fixed and an adequate number of popholes is required based on the size of the houses and number of animals [40]. In traditional free-range, such as the label rouge systems in France, stipulates a minimum slaughter age of 81 days and no more than 4800 broilers can be produced per flock [36,40]. In Brazil, the "caipira" production system (another example of national labeling program) specifies regulations for free-range systems. The amount of free-range space is subjective as there are no regulations requiring a standard stocking density [37]. Depending on the amount of time broilers spend outside, it could be argued that free-range are no different than conventional if the free-range chickens never leave the housing structure [37]. In free-range poultry systems, runs are often not well managed and animals may not feel safe in it. Outdoor runs could be made more attractive by offering different kinds of shelter [4,6]. A free-range area contributes to the welfare of laying hens. Studies that looked at the relation between the use of a free-range area and the degree of feather pecking damage, found that if a higher proportion of a flock uses the free range area, then significantly less feather pecking damage is seen [41–43]. A higher proportion of hens using the free range area can be achieved by providing shelter [41]. This can be artificial structures or natural, for example trees or bushes. It was found a relation between tree cover and injurious feather-pecking: less feather pecking was seen in the case of more tree cover [44]. Besides an advantage for animal welfare, a higher degree of woody cover in the free-range area seems to be related to less avian influenza risk for birds in the free range area [43]. The presence of trees can help chicken feel more secure from predators and more sheltered from sun and the elements so they can venture further away from the huts and eat more forage [45]. Another benefit of tree and forage cover is a better distribution of hens across the range area, which may reduce the risk of parasitic contamination [46].

Finally, a better distribution of hens may prevent local accumulation of nitrogen and phosphate [47]. It has been shown that range enrichment is not only beneficial for the animal behavior [48], but can also be economically advantageous (e.g., reduction of animal mortality rate), resulting in a win–win situation for poultry welfare and production [6]. According to Cobanoglu et al. [49], free-range producers commonly utilize the same fast-growing broilers as conventional systems; however, the suitability of these birds may not be adequate since conventional broilers do not acclimate themselves well with outdoor conditions [39].

3.2. Organic System

Organic refers to the way agricultural and livestock products are reared and processed, which involves avoiding agrichemicals such as synthetic fertilizers and pesticides. Given the current demand for organic foods, organic poultry production has become a growing segment of the poultry industry. Organic poultry have access to pasture [6,39], a nutrient source that has not been completely investigated for use in poultry. Organic poultry production focuses particularly on animal health and wellbeing, good environmental practices, and product quality and focuses less on economic concerns, such as reducing costs and maximizing production [37,50].

Overall, the EU regulations (No 889/2008) [51] for organic production of poultry specify that land should be free of synthetic fertilizers, pesticides and herbicides for a specific time. The dietary supplementation is subject to some specific requirements and synthetic ingredients and GMOs in feed are regularly forbidden [36]. In some European countries and in the U.S., some limiting synthetic amino acids are still permitted at low levels if organic sources are not available to farmers. However, in the EU, the feed ingredients must be organically produced, if not the meat or eggs will not be labeled as organic. In the U.S., organic labeling is available at three different levels based on the amount of organic feed supplemented to broilers: "100% organic broiler"; "organic broiler", meaning that at least 95% of the ingredients were produced organically; and "made with organic ingredients", meaning that birds consumed at least 70% of organic ingredients in feed [36,52]. The U.S. regulation does not specify indoor or outdoor stocking density, but certification agencies look for a maximum of 8 birds/m^2 indoors [37].

In European countries, even though the law is the same for all members, there are specific local regulations and labeling programs related to the organic poultry production. In particular, the maximum flocks size cannot exceed 4800 broilers and a maximum stocking rate of 10 birds/m^2 is obligatory. Further, if mobile housing is used, the stocking density can be increased to 16 broilers/m^2, but the maximum flocks size cannot be modified [36,52]. Poultry must be able to walk around the outdoor runs at stocking rate of 4 m^2 per subject. Regarding the minimum age at slaughter, the starting point is the 81 days specified in the European regulation (No 889/2008) [51]. Under organic production systems, antibiotics and drugs are substituted by alternative natural products [53] and birds are raised for longer periods (>81 days) until slaughter.

4. Influence of Feeding Forage on Poultry Egg and Meat Quality

Poultry diets are commonly corn–soybean meal-based displaying a high energy concentration and low fiber levels. Nevertheless, many feedstuffs having high fiber content have been commonly included in poultry diet, especially in extensive poultry production systems. However, depending on the solubility levels and concentrations, fiber in diets influences poultry performance [54–56].

Outdoor systems are sustainable alternatives for poultry production where laying hens are able to ingest nutrients from pasture and forage, minimizing the intake of commercial feed. Poultry can eat forage, pebbles, weeds, crop seeds, earthworms and insects in the paddock [57]. It was also reported that an outdoor system could benefit the environment because of increased nutrients circulation within the system [58].

Laying hens and broiler chickens given access to pasture may meet various nutrient needs through foraging [7,59]. It was reported that laying hens having access to grass resulted in a 20% reduction in feed consumption and increased egg production compared with hens raised under conventional conditions [59,60]. Moreover, hens reared on alfalfa or clover need significantly less feed protein than confined hens [7]. In particular, alfalfa forage can supply carotene, vitamins and other nutrients, especially high amount of protein up to 19% as dry matter basis [61,62]. In addition, it was found that organically reared broilers may overcome growth impairments associated with methionine deficiency through foraging [63]. The primary benefit of forage consumption is that plant matter is typically high in both vitamins and minerals; moreover, forages contain components such as fiber, protein, energy, and other compounds such as carotenoids and n-3 fatty acids having important metabolic functions in all animals, including poultry [64]. Pasture intake by poultry may acts as a form of nutritional insurance and pasturing poultry and giving them access to high-quality forages will help in balancing out any deficiencies. Moreover, forages can provide supplemental minerals, and the calcium found in plants such as alfalfa is highly bioavailable. Poultry digestive system is able to utilize calcium from forages as efficiently as calcium from more common sources such as limestone or oyster shell [61]. It was reported that broilers and laying hens fed low-protein diets increased their intake of forages compared to flocks fed a ration with adequate protein concentration [58,65]. Similarly, poultry nutritionists have found that forages consumption is inversely tied to protein levels, since a higher protein content in the diet can result in a lower amount of plant matter consumed on pasture [63]. Further, poultry are able to utilize most of amino acids consuming forages, resulting in a significant level of methionine and lysine digested (88% and 79%, respectively) [7,59]. On the other hand, forages are low in energy, but these plants can contribute to the overall energy requirements of poultry. In this regard, Rivera-Ferre et al. [66] reported that poultry raised on pasture obtained only 3% of their energy requirement from forages. Although grains are noticeably the most important energy source in poultry diet along with oils, even the low amounts of energy supplied by forages are vital when feed prices increase [59].

Research is beginning to investigate claims of particular nutritive characteristics of eggs and meat from pastured birds. It was assessed that the poultry products from grass-fed flocks tend to have less total cholesterol and more vitamins A and E, as well as high omega-3, and an improved omega-3 to omega-6 ratio [67]. It was found that egg cholesterol decreased as alfalfa meal increased in a laying hen diet [62]. Studies demonstrated that hens with access to high-quality pasture had eggs with at least twice as much vitamins A and E and omega-3s, compared to hens having no access to different pastures [68–71]. Furthermore, it is well established that a large range of forages, such as alfalfa, perennial ryegrass, red clover, and grass meals, although containing significant fiber levels, are valuable sources of xanthophylls and can be successfully used in diets as natural pigmenting agents [72]. Available studies have reported increasing levels of n-3 fatty acids in meat from pasture-raised broilers as well as higher levels of vitamin E and other nutrients [34,35,71].

The evaluation of including dehydrated leguminous-based forages in common poultry diets under intensive production systems on growth performance and meat quality remains to be more deeply investigated. Leguminous-based forages may be considered as a source of fiber and protein for broilers, and forages are also a source of natural antioxidants [73]. Antioxidant supplementation of feed is an efficient method for increasing poultry meat oxidative stability [74]. Moreover, β-carotene, a pro-vitamin A compound abundant in forages, is the predominant carotenoid in meat and meat products [75]. Feeding of dehydrated leguminous-based forage vitamin E homologs and β-carotene for broiler meat oxidative stability remains, however, to be established [73]. It has been previously found that including dehydrated leguminous forages in broiler diet contributes to decrease cholesterol levels of meat [76]. Conversely, it was demonstrated that the consumption of a dehydrated leguminous-based forage, offered free choice and ad libitum, by broilers of a fast-growing genotype exploited under an intensive production system had no effect on broiler performance, on the profiles of vitamin E homologs and on the cholesterol content of meat [73]. However, the intake of the dehydrated forage had a major influence in the fatty acid profile of broiler meat, resulting the levels of n-3 long-chain PUFA (EPA, DPA, and DHA) in breast meat significantly higher when animals consumed the leguminous biomass. Further, it was reported that feeding of citrus pulp or dehydrated pasture at 10% levels changed broiler meat fatty acid profiles, depressing MUFA and increasing the predominance of n-6 and n-3 PUFA [8]; thus, these findings suggested that dehydrated pasture may be an interesting supplement to add in broiler diet to enhance meat lipid quality.

Several antinutritional substances (such as saponins, phenolic compounds and protease inhibitors) influencing negatively the performance of poultry fed forage-diets have been reported. For example, based on the findings of many studies, it was found that dehydrated alfalfa may contain some antinutritional factors (saponins and isoflavones) reducing the productive performance of both broilers and laying hens [76–80]. It was also assessed that poultry are more sensitive to dietary alfalfa saponins than other monogastric species [81]. Levels up to 20% alfalfa meal in broiler rations resulted in growth rate depression due to the saponin content [81]. Further, tannic acid was also found to reduce the metabolizable energy of the broilers diet, depressing the nitrogen retention by chicks [82]. Finally, in a series of studies with geese, turkeys, quails and chickens fed with high-saponin and low-saponin alfalfa meal, at levels ranging 1–20% of diet, it was found that the only discrimination between the two alfalfa types was observed in geese fed 20% alfalfa [23].

5. Conclusions and Future Outlook

The improvement of poultry production systems may be essential to produce quality animal products; thus, it is crucial that production systems are compliant to the needs of populations being associated with animal production in both developing and developed countries. The use forage as the advantageous alternative to the conventional feeds may be an input to increase the forages utilization in poultry production systems. In alternative poultry production systems, both organic and free-range, it is imperative to improve the outdoor area utilization and to optimize forage intake, which still requires further investigation in poultry. Poultry production would benefit from feeding forage as substitute to conventional cereals and oil seeds to reduce the dependence on these feedstuffs, especially in particular areas of the world. Moreover, there are many forage species as alternative protein sources for livestock species, thus forages could provide the basis for most animal production systems for the near future.

Author Contributions: All the authors equally contributed and commented on early and final version of manuscript.

Acknowledgments: No funds or grants were received in support of our paper.

Conflicts of Interest: The authors declare no conflict of interest.

References

1. Siderer, Y.; Maquet, A.; Anklam, E. Need for research to support consumer confidence in the growing organic food market. *Trends Food Sci. Technol.* **2005**, *16*, 332–343. [CrossRef]
2. Miao, Z.H.; Glatz, P.C.; Ru, Y.J. Free-range poultry production—A review. *Asian-Aust. J. Anim. Sci.* **2005**, *18*, 113–132. [CrossRef]
3. Rigby, D.; Cáceres, D. Organic farming and the sustainability of agricultural systems. *Agric. Syst.* **2001**, *68*, 21–40. [CrossRef]
4. Berg, C. Health and welfare in organic poultry production. *Acta Vet. Scand.* **2002**, *43*, S37. [CrossRef]
5. Castellini, C.; Bastianoni, S.; Granai, C.; Dal Bosco, A.; Brunetti, M. Sustainability of poultry production using the emergy approach: Comparison of conventional and organic rearing systems. *Agric. Ecosyst. Environ.* **2006**, *114*, 343–350. [CrossRef]

6. Dal Bosco, A.; Mugnai, C.; Rosati, A.; Paoletti, A.; Caporali, S.; Castellini, C. Effect of range enrichment on performance, behavior, and forage intake of free-range chickens. *J. Appl. Poult. Res.* **2014**, *23*, 137–145. [CrossRef]

7. Buchanan, N.P.; Hott, J.M.; Kimbler, L.B.; Moritz, J.S. Nutrient composition and digestibility of organic broiler diets and pasture forages. *J. Appl. Poult. Res.* **2007**, *16*, 13–21. [CrossRef]

8. Mourão, J.L.; Pinheiro, V.M.; Prates, J.A.M.; Bessa, R.J.B.; Ferreira, L.M.A.; Fontes, C.M.G.A.; Ponte, P.I.P. Effect of dietary dehydrated pasture and citrus pulp on the performance and meat quality of broiler chickens. *Poult. Sci.* **2008**, *87*, 733–743. [CrossRef] [PubMed]

9. Abouelezz, F.M.K.; Sarmiento-Franco, L.; Santos-Ricalde, R.; Solorio-Sanchez, F. Outdoor egg production using local forages in the tropics. *World's Poult. Sci. J.* **2012**, *68*, 679–692. [CrossRef]

10. Wu, G.; Fanzo, J.; Miller, D.D.; Pingali, P.; Post, M.; Steiner, J.L.; Thalacker-Mercer, A.E. Production and supply of high-quality food protein for human consumption: Sustainability, challenges, and innovations. *Ann. N. Y. Acad. Sci.* **2014**, *1321*, 1–19. [CrossRef] [PubMed]

11. Gerber, P.J.; Mottet, A.; Opio, C.I.; Falcucci, A.; Teillard, F. Environmental impacts of beef production: Review of challenges and perspectives for durability. *Meat Sci.* **2015**, *109*, 2–12. [CrossRef] [PubMed]

12. Guyader, J.; Janzen, H.H.; Kroebel, R.; Beauchemin, K.A. Forage use to improve environmental sustainability of ruminant production. *J. Anim. Sci.* **2016**, *94*, 3147–3158. [CrossRef] [PubMed]

13. Chaudhry, A.S. Forage based animal production systems and sustainability, an invited keynote. *Rev. Bras. Zootec.* **2008**, *37*, 78–84. [CrossRef]

14. Dale, A.J.; Laidlaw, A.S.; Frost, J.P.; Bailey, J.; Mayne, C.S. *Opportunities to Improve Efficiency of Use of Animal Manures with Low Input, Alternative Forages*; Occasional Symposium-British Grassland Society: Berks, UK, 2007; p. 205.

15. Lüscher, A.; Mueller-Harvey, I.; Soussana, J.F.; Rees, R.M.; Peyraud, J.L. Potential of legume-based grassland–livestock systems in Europe: A review. *Grass Forage Sci.* **2014**, *69*, 206–228. [CrossRef] [PubMed]

16. INRA. *Alimentation des Bovins, ovins et Caprins. Besoins des animaux. Valeur des Aliments*; Feeding of Cattle, Sheep and Goats. Animal Needs. Feed Value. Tables INRA: Editions Quae: Paris, France, 2007.

17. Waghorn, G. Beneficial and detrimental effects of dietary condensed tannins for sustainable sheep and goat production—Progress and challenges. *Anim. Feed Sci. Technol.* **2007**, *147*, 116–139. [CrossRef]

18. Waghorn, G.C.; McNabb, W.C. Consequences of plant phenolic compounds for productivity and health of ruminants. *Proc. Nutr. Soc.* **2003**, *62*, 383–392. [CrossRef] [PubMed]

19. Martens, S.D.; Tiemann, T.T.; Bindelle, J.; Peters, M.; Lascano, C.E. Alternative plant protein sources for pigs and chickens in the tropics-nutritional value and constraints: A review. *J. Agric. Rural Dev. Trop. Subtrop.* **2012**, *113*, 101–123.

20. Grant, G.; More, L.J.; McKenzie, N.H.; Dorward, P.M.; Buchan, W.C.; Telek, L.; Pusztai, A. Nutritional and haemagglutination properties of several tropical seeds. *J. Agric. Sci.* **1995**, *124*, 437–445. [CrossRef]

21. Tufarelli, V.; Casalino, E.; D'Alessandro, A.G.; Laudadio, V. Dietary Phenolic Compounds: Biochemistry, Metabolism and Significance in Animal and Human Health. *Curr. Drug Metab.* **2017**, *18*, 905–913. [CrossRef] [PubMed]

22. Brum, K.B.; Haraguchi, M.; Garutti, M.B.; Nóbrega, F.N.; Rosa, B.; Fioravanti, M.C.S. Steroidal saponin concentrations in Brachiaria decumbens and B. brizantha at different developmental stages. *Ciência Rural* **2009**, *39*, 279–281. [CrossRef]

23. Cheeke, P.R.; Powley, J.S.; Nakaue, H.S.; Arscott, G.H. Feed preferences of poultry fed alfalfa meal, high and low saponins alfalfa and quinine sulfate. *Proc. West. Sect. Am. Soc. Anim. Sci.* **1981**, *32*, 426–427.

24. Pedersen, M.W.; Berrang, B.; Wall, M.E.; Davis, K.H. Modification of saponin characteristics of alfalfa by selection 1. *Crop Sci.* **1973**, *13*, 731–735. [CrossRef]

25. Acamovic, T.; Cowieson, A.; Gilbert, C.E. Lupins in poultry nutrition. Wild and cultivated lupins from the Tropics to the Poles. In Proceedings of the 10th International Lupin Conference, Laugarvatn, Iceland, 19–24 June 2002; pp. 314–318.

26. Jeroch, H.; Flachowsky, G.; Weissbach, F. *Futtermittelkunde*; G. Fischer: Jena, Germany; Stuttgart, Germany, 1993.

27. Barry, T.N.; McNeill, D.M.; McNabb, W.C. Plant secondary compounds; their impact on nutritive value and upon animal production. In Proceedings of the XIX International Grassland Congress, Sao Paulo, Brazil, 11–21 February 2001; pp. 445–452.

28. Akande, K.E.; Doma, U.D.; Agu, H.O.; Adamu, H.M. Major antinutrients found in plant protein sources: Their effect on nutrition. *Pak. J. Nutr.* **2010**, *9*, 827–832. [CrossRef]

29. Laudadio, V.; Tufarelli, V. Dehulled-micronised lupin (*Lupinus albus* L. cv. Multitalia) as the main protein source for broilers: Influence on growth performance, carcass traits and meat fatty acid composition. *J. Sci. Food Agric.* **2011**, *91*, 2081–2087. [CrossRef] [PubMed]

30. Laudadio, V.; Nahashon, S.N.; Tufarelli, V. Growth performance and carcass characteristics of guinea fowl broilers fed micronized-dehulled pea (*Pisum sativum* L.) as a substitute for soybean meal. *Poult. Sci.* **2012**, *91*, 2988–2996. [CrossRef] [PubMed]

31. Vadivel, V.; Janardhanan, K. Nutritional and antinutritional characteristics of seven South Indian wild legumes. *Plant Foods Hum. Nutr.* **2005**, *60*, 69–75. [CrossRef] [PubMed]

32. Khattab, R.Y.; Arntfield, S.D. Nutritional quality of legume seeds as affected by some physical treatments 2. Antinutritional factors. *LWT-Food Sci. Technol.* **2009**, *42*, 1113–1118. [CrossRef]

33. Castellini, C.; Mugnai, C.; Dal Bosco, A. Effect of organic production system on broiler carcass and meat quality. *Meat Sci.* **2002**, *60*, 219–225. [CrossRef]

34. Fanatico, A.C.; Pillai, P.B.; Emmert, J.L.; Owens, C.M. Meat quality of slow-and fast-growing chicken genotypes fed low-nutrient or standard diets and raised indoors or with outdoor access. *Poult. Sci.* **2007**, *86*, 2245–2255. [CrossRef] [PubMed]

35. Castellini, C.; Berri, C.; Le Bihan-Duval, E.; Martino, G. Quality attributes and consumer perception of organic and free range poultry meat. *World's Poult. Sci. J.* **2008**, *64*, 500–512. [CrossRef]

36. Almeida, G.F. Use of forage and plant supplements in organic and free range broiler systems: Implications for production and parasite infections. PhD Thesis, Aarhus University, Denmark, 2012.

37. Fanatico, A.C.; Owens, C.M.; Emmert, J.L. Organic poultry production in the United States: Broilers. *J. Appl. Poult. Res.* **2009**, *18*, 355–366. [CrossRef]

38. Martinez-Perez, M.; Sarmiento-Franco, L.; Santos-Ricalde, R.H.; Sandoval-Castro, C.A. Poultry meat production in free-range systems: Perspectives for tropical areas. *World's Poult. Sci. J.* **2017**, *73*, 309–320. [CrossRef]

39. Husak, R.L.; Sebranek, J.G.; Bregendahl, K. A survey of commercially available broilers marketed as organic, free-range, and conventional broilers for cooked meat yields, meat composition, and relative value. *Poult. Sci.* **2008**, *87*, 2367–2376. [CrossRef] [PubMed]

40. Council Regulation (EC). No 1234/2007 of 22 October 2007 establishing a common organisation of agriculture markets and on specific provisions for certain agricultural products. *Off. J. Eur. Union* **2007**, *299*, 1–149.

41. Smith, D.P.; Northcutt, J.K.; Steinberg, E.L. Meat quality and sensory attributes of a conventional and a Label Rouge-type broiler strain obtained at retail. *Poult. Sci.* **2012**, *91*, 1489–1495. [CrossRef] [PubMed]

42. Bestman, M.W.P.; Wagenaar, J.P. Farm level factors associated with feather pecking damage in organic laying hens. *Livest. Prod. Sci.* **2003**, *80*, 133–140. [CrossRef]

43. Lambton, S.L.; Knowles, T.G.; Yorke, C.; Nicol, C.J. The risk factors affecting the development of gentle and severe feather pecking in loose housed laying hens. *Appl. Anim. Behav. Sci.* **2010**, *123*, 32–42. [CrossRef]

44. Bestman, M.; de Jong, W.; Wagenaar, J.; Weerts, T. Presence of avian influenza risk birds in and around poultry free-range areas in relation to range vegetation and openness of surrounding landscape. *Agrofor. Syst.* **2017**. [CrossRef]

45. Bright, A.; Gill, R.; Willings, T.H. Tree cover and injurious feather-pecking in commercial flocks of free-range laying hens: A follow up. *Anim. Welf.* **2016**, *25*, 1–5. [CrossRef]

46. Sossidou, E.N.; Dal Bosco, A.; Castellini, C.; Grashorn, M.A. Effects of pasture management on poultry welfare and meat quality in organic poultry production systems. *World's Poult. Sci. J.* **2015**, *71*, 375–384. [CrossRef]

47. Bray, T.S.; Lancaster, M. BThe parasitic status of land used by free-range hens. *Br. Poult. Sci.* **1992**, *33*, 1119–1120.

48. Dekker, S.E.M.; Aarnink, A.J.A.; de Boer, I.J.M.; Groot Koerkamp, P.W.G. Total loss and distribution of nitrogen and phosphorus in the outdoor run of organic laying hens. *Br. Poult. Sci.* **2012**, *53*, 731–740. [CrossRef] [PubMed]

49. Larsen, H.; Cronin, G.; Smith, C.L.; Hemsworth, P.; Rault, J.L. Behaviour of free-range laying hens in distinct outdoor environments. *Anim. Welf.* **2017**, *26*, 255–264. [CrossRef]

50. Cobanoglu, F.; Kucukyilmaz, K.; Cinar, M.; Bozkurt, M.; Catli, A.U.; Bintas, E. Comparing the profitability of organic and conventional broiler production. *Rev. Bras. Ciên. Avíc.* **2014**, *16*, 89–95. [CrossRef]
51. Commission Regulation (EC) No 889/2008 of 5 September 2008 laying down detailed rules for the implementation of Council Regulation (EC) No 834/2007 on organic production and labelling of organic products with regard to organic production, labelling and control. *Off. J. Eur. Union* **2008**, *280*, 1–84.
52. USDA. National organic program; amendment to the national list of allowed and prohibited substances (livestock). *Fed. Reg.* **2012**, *77*, 57985–57990.
53. Anon. *Report from the Danish Poultry Council 2010*; Danish Poultry Council: Copenhagen, Denmark, 2012.
54. Khan, R.U.; Naz, S.; Nikousefat, Z.; Tufarelli, V.; Javdani, M.; Qureshi, M.S.; Laudadio, V. Potential applications of ginger (*Zingiber officinale*) in poultry diets. *World's Poult. Sci. J.* **2012**, *68*, 245–252. [CrossRef]
55. Mathlouthi, N.; Mallet, S.; Saulnier, L.; Quemener, B.; Larbier, M. Effects of xylanase and β-glucanase addition on performance, nutrient digestibility, and physico-chemical conditions in the small intestine contents and caecal microflora of broiler chickens fed a wheat and barley-based diet. *Anim. Res.* **2002**, *51*, 395–406. [CrossRef]
56. Singh, A.K.; Berrocoso, J.D.; Dersjant-Li, Y.; Awati, A.; Jha. R. Effect of a combination of xylanase, amylase and protease on growth performance of broilers fed low and high fiber diets. *Anim. Feed Sci. Technol.* **2017**, *232*, 16–20. [CrossRef]
57. Tufarelli, V.; Dario, M.; Laudadio, V. Effect of xylanase supplementation and particle-size on performance of guinea fowl broilers fed wheat-based diets. *Int. J. Poult. Sci.* **2007**, *4*, 302–307. [CrossRef]
58. Horsted, K.; Hermansen, J.; Ranvig, H. Crop content in nutrient-restricted versus non-restricted organic laying hens with access to different forage vegetations. *Br. Poult. Sci.* **2007**, *48*, 177–184. [CrossRef] [PubMed]
59. Spencer, T. *Pastured Poultry Nutrition and Forages*; ATTRA: Melbourne, Australia, 2013; Available online: https://attra.ncat.org/attra-pub/summaries/summary.php?pub=452 (accessed on 13 November 2017).
60. Buckner, G.D.; Insko, W.M., Jr.; Henry, A.H. Influences of spring bluegrass and mature bluegrass pastures on laying hens and on the eggs produced. *Poult. Sci.* **1945**, *24*, 446–450. [CrossRef]
61. Blair, R. *Nutrition and Feeding of Organic Poultry*; CAB International Publishing: Oxfordshire, UK, 2008.
62. Laudadio, V.; Ceci, E.; Lastella, N.M.B.; Introna, M.; Tufarelli, V. Low-fiber alfalfa (*Medicago sativa* L.) meal in the laying hen diet: Effects on productive traits and egg quality. *Poult. Sci.* **2014**, *93*, 1868–1874. [CrossRef] [PubMed]
63. Moritz, J.S.; Parsons, A.S.; Buchanan, N.P.; Baker, N.J.; Jaczynski, J.; Gekara, O.J.; Bryan, W.B. Synthetic methionine and feed restriction effects on performance and meat quality of organically reared broiler chickens. *J. Appl. Poult. Res.* **2005**, *14*, 521–535. [CrossRef]
64. Moyle, J.R.; Arsi, K.; Woo-Ming, A.; Arambel, H.; Fanatico, A.; Blore, P.J.; Clark, F.D.; Donoghue, D.J.; Donoghue, A.M. Growth performance of fast-growing broilers reared under different types of production systems with outdoor access: Implications for organic and alternative production systems. *J. Appl. Poult. Res.* **2014**, *23*, 212–220. [CrossRef]
65. Eriksson, M. Protein supply in organic broiler production using fast-growing hybrids. PhD Dissertation, Uppsala, Sweden, 2010. Available online: http://pub.epsilon.slu.se/2362/1/eriksson_m_101008.pdf (accessed on 6 May 2017).
66. Rivera-Ferre, M.G.; Lantinga, E.A.; Kwakkel, R.P. Herbage intake and use of outdoor area by organic broilers: Effects of vegetation type and shelter addition. *NJAS-Wagen. J. Life Sci.* **2007**, *54*, 279–291. [CrossRef]
67. Anderson, K.E. Comparison of fatty acid, cholesterol, and vitamin A and E composition in eggs from hens housed in conventional cage and range production facilities. *Poult. Sci.* **2011**, *90*, 1600–1608. [CrossRef] [PubMed]
68. Mugnai, C.; Dal Bosco, A.; Castellini, C. Effect of rearing system and season on the performance and egg characteristics of Ancona laying hens. *Ital. J. Anim. Sci.* **2009**, *8*, 175–188. [CrossRef]
69. Karsten, H.D.; Patterson, P.H.; Stout, R.; Crews, G. Vitamins A, E and fatty acid composition of the eggs of caged hens and pastured hens. *Renew. Agric. Food Syst.* **2010**, *25*, 45–54. [CrossRef]
70. Holt, P.S.; Davies, R.H.; Dewulf, J.; Gast, R.K.; Huwe, J.K.; Jones, D.R.; Waltman, D.; Willian, K.R. The impact of different housing systems on egg safety and quality. *Poult. Sci.* **2011**, *90*, 251–262. [CrossRef] [PubMed]
71. Dal Bosco, A.; Mugnai, C.; Mattioli, S.; Rosati, A.; Ruggeri, S. Ranucci, D.; Castellini, C. Transfer of bioactive compounds from pasture to meat in organic free-range chickens. *Poult. Sci.* **2016**, *95*, 2464–2471. [CrossRef] [PubMed]

72. Grigorova, S.; Abadjieva, D.; Gjorgovska, N. Influence of natural sources of biologically active substances on livestock and poultry reproduction. *Iran. J. Appl. Anim. Sci.* **2017**, *7*, 189–195.

73. Ponte, P.I.; Prates, J.A.; Crespo, J.P.; Crespo, D.G.; Mourão, J.L.; Alves, S.P.; Bessa, R.J.; Chaveiro-Soares, M.A.; Ferreira, L.M.; Fontes, C.M. Improving the lipid nutritive value of poultry meat through the incorporation of dehydrated leguminous-based forage in the diet for broiler chicks. *Poult. Sci.* **2008**, *87*, 1578–1594. [CrossRef] [PubMed]

74. Tufarelli, V.; Laudadio, V.; Casalino, E. An extra-virgin olive oil rich in polyphenolic compounds has antioxidant effects in meat-type broiler chickens. *Environ. Sci. Pollut. Res.* **2016**, *23*, 6197–6204. [CrossRef] [PubMed]

75. Mortensen, A.; Skibsted, L.H. Kinetics and mechanism of the primary steps of degradation of carotenoids by acid in homogeneous solution. *J. Agric. Food Chem.* **2000**, *48*, 279–286. [CrossRef] [PubMed]

76. Ponte, P.I.P.; Mendes, I.; Quaresma, M.; Aguiar, M.N.M.; Lemos, J.P.C.; Ferreira, L.M.A.; Soares, M.A.C.; Alfaia, C.M.; Prates, J.A.M.; Fontes, C.M.G.A. Cholesterol levels and sensory characteristics of meat from broilers consuming moderate to high levels of alfalfa. *Poult. Sci.* **2004**, *83*, 810–814. [CrossRef] [PubMed]

77. Heywang, B.W.; Thompson, C.R.; Kemmerer, A.R. Effect of alfalfa saponins on laying hens. *Poult. Sci.* **1959**, *38*, 968–971. [CrossRef]

78. Reddy, B.S.V. Feeding Value of Forages in Poultry. PhD Thesis, Punjab Agricultural University, Ludhiana, India, 1979.

79. Sim, J.S.; Kitts, W.D.; Bragg, D.B. Effect of dietary saponin on egg cholesterol level and laying hen performance. *Can. J. Anim. Sci.* **1984**, *64*, 977–984. [CrossRef]

80. Kocaoğlu Güçlü, B.; Işcan, K.M.; Uyanik, F.; Eren, M.; Can Ağca, A. Effect of alfalfa meal in diets of laying quails on performance, egg quality and some serum parameters. *Arch. Anim. Nutr.* **2004**, *58*, 255–263. [CrossRef] [PubMed]

81. Sen, S.; Makkar, H.P.; Becker, K. Alfalfa saponins and their implication in animal nutrition. *J. Agric. Food Chem.* **1998**, *46*, 131–140. [CrossRef] [PubMed]

82. Vohra, P.; Kratzer, F.H.; Joslyn, M.A. The growth depressing and toxic effects of tannins to chicks. *Poult. Sci.* **1966**, *45*, 135–142. [CrossRef] [PubMed]

agriculture

MDPI

Review

The Efficacy of High-Protein Tropical Forages as Alternative Protein Sourcesfor Chickens: A Review

Sameh A. Abdelnour [1], Mohamed E. Abd El-Hack [2],* and Marco Ragni [3]

[1] Department of Animal Production, Faculty of Agriculture, Zagazig University, Zagazig 44511, Egypt; samehtimor86@gmail.com
[2] Department of Poultry, Faculty of Agriculture, Zagazig University, Zagazig 44511, Egypt
[3] Department of Agricultural and Environmental Science, University of Bari Aldo Moro, 70121 Bari, Italy; marco.ragni@uniba.it
* Correspondence: dr.mohamed.e.abdalhaq@gmail.com or m.ezzat@zu.edu.eg; Tel.: +20-10-668-6449

Received: 4 June 2018; Accepted: 14 June 2018; Published: 20 June 2018

Abstract: Smallholders of poultry production systems in developing countries are commonly found in rural, resource-poor areas, and often face food insecurity. The main constraints for smallholders in poultry production in rural, resource-poor areas are the shortage of available commercial dietary protein and the high cost of commercial diets. The beneficial effects of legume and forage cultivation are economic, through providing protein for animals, and ecological, such as soil amendment, nitrogen fixation, and stripping control which participate to increase cropping efficiency. The potential nutritive value of a wide range of forages and grain legumes is presented and discussed. The impacts of dietary protein, fiber, and secondary metabolites in plant content, as as well as their consequences on feed efficiency, animal performance, and digestion processes are enclosed in this review. Lastly, approaches to reduce the anti-nutritional factors of the secondary metabolites of plants are explained.

Keywords: tropical forages; chicken; alternative protein; anti-nutritional factors

1. Introduction

In developing countries, there is an increasing demand for animal protein, which is principally poultry products [1]. About 20% of the world's population is considered smallholders with livestock, and they have a great opportunity to improve income and raise their sustenance through the development of the livestock chain [2]. One of the most important concerns of livestock smallholders is to get good quality rations of energy, protein, amino acids, minerals, and vitamins to ensure suitably high productivity of their animals. It is recognized that soybean meal is often used as material feed, as it has high contents of amino acid profiles and energy in livestock and poultry rations. Global production of soybean reached 366 million tons in 2016, and the United States has the highest rate of soybean production (117.2), followed by Brazil (96.3) and Argentina (58.8), according to FAOSTAT [3]. After oil is extracted from soybeans, the residuals are expressed as a soybean meal, and it is mainly used in poultry production systems and for other livestock animals as ingredients in their diet. Globally, there is little quantity of soybean meal for smallholders. Additionally, the high cost of feed concentration for livestock is progressive [1]. Consequently, to meet the nourishment requirements for livestock or poultry, alternative low-cost feed resources must be recognized [4].

Tropical and subtropical areas are characterized by raised ambient temperatures and water deficiencies, in addition to tropical soils, which suffer from a lack of nitrogen. Thus, production of protein-rich material in the diets of small animals either involves the input of nitrogen fertilizer to gramineous crops, or the use of legumes either as the source of the supplement itself or as part of a rotation. Here, we have focused on the use of legume crops as alternatives [5]. Based on the large diversity of legumes in humid and sub-humid areas, about 650 genera of legumes and

18,000 certain species have been identified [6]. The International Livestock Research Institute [7], the Centro Internacional de Agricultura Tropical [8], in addition to the Australian Tropical Crops and Forages Collection and the collection of CENARGEN-EMBRAPA, have collected and evaluated many of tropical forages and crops within gene banks.

This review paper aims to distinguish alternative resources of primary feed for poultry, and identifies the options for improving smallholder production of monogastric animals in the tropics in terms of their protein needs and forage supply.

2. Nutrient Utilization Chickens

Chickens have a simple stomach which is expressed as monogastric. On the contrary, ruminants have several parts to their stomachs, which are called complex stomachs. This function qualifies these animals to digest fiber. Although clear dissimilarities between digestive systems have been observed between birds and monogastric mammals, they have commensurate general feed digestion patterns paralleled with polygastrics. It is known that feed is digested by enzymes and acid in the stomach, and soluble constituents are absorbed mainly via the epithelial cells in the small intestine. Indigestible compounds, such as non-starch polysaccharides, proteins, and resistant starches subjected to Maillard reactions, as well as some fiber bound proteins and tannins, reach the ceca and cecum in poultry, where, together with endogenous secretions, they are fermented by the gut microflora. The end products of the fermentation in hindgut are short-chain fatty acids (SCFA), which are an equally important energy source for the microbiome. In chickens and ostriches, evidence has been obtained suggesting that SCFA can provide up to 8 to 75% of their energy requirements from fermentation in the ceca [9]. A small amount of microbial amino acids, which have been synthesized by the microbiome in the gut, can be soaked up in the intestines. Thus, stomach enzymes can digest feed protein to be absorbed in the small intestine. An animal's specific amino acid requirements should be available in the feed protein as amino acids. The ideal protein requirement for an animal depends on several factors such as type of production, stage of growth, product, season, and composition of body tissue, which need to be taken into account in order for it to be sufficient for maintenance and growth [10].

3. Production System Limitations in the Tropics

It is common for deficiency of essential amino acids to be present in smallholder systems characterized monogastric production, especially diets involving cereal grain, a blend of them (e.g., rice bran, rice, sorghum, ormaize) or cassava. Farmers often do not know or comprehend the nutritional value of these alternative feeds, nor do they know their animals' feed requirements. Besides, the nutritional quality of these alternatives may be low owing to fiber-bound nitrogen [11], and compounds such as tannins and trypsin inhibitors may be related to the inhibition of enzyme function or they may bind to proteins, diminishing their digestibility. The limitation in choices of feed in developing countries makes smallholders search for alternative specific feeds for their poultry, particularly those that increase productivity of poultry. The enrichment of low-protein diets with synthetic amino acids such as lysine, reduces Nexcretion and accelerates growth rates [4,12], and it is actually significant to compensate amino acid deficiency in poultry diets.

Application of this option in commercial production systems is obtainable, but it is rarely favorable or available for most smallholders. Thus, they avoid feeding animals with high protein diets, as the excess degrades to uric acid or urea for excretion. This phenomenon makes the animal lose a large amount of energy, and moreover it causes harmful effects on the environment [13]. Smallholder farms are seeking rapid growth rates for their chickens; however, with local forages this may be less obtainable and profitable.

4. Tropical Forages as a Protein Source

As a matter of choice, it is preferable to obtain feed from crop derivatives which are part of environmentally sustainable farming systems in the region. From this point of view, biomass productivity per unit of solar energy should be optimized, inputs of agro-chemicals reduced, and soil fertility and biodiversity sustained [4]. However, most of these requirements are scarcely met at the same time, so it has been suggested that tropical forages as feed for monogastric animals can play a role in improving the sustainability of animal production within farming systems through [14,15]:

(1). Increased production of biomass in environments where it is not suitable for other crops;
(2). Feed with high protein and amino acid profiles, particularly, sulfur amino acids, methionine, and lysine, which for monogastrics sufficiently balances the constraints of cereal proteins [leaf and grain];
(3). High levels of minerals and vitamins compared to conventional energy-based feed requirements.

5. Nutritional Value and Impact on Animal Performance

Tropical forage plants have shown to widely differ in crude protein content, and have almost reach 36% of dry matter (DM)in some forages, which is parallel to soybean grain. The plants analyzed for chickens are categorized by lowering the proportion of the sulphur-containing methionine and cysteine, when comparing the amino acid profiles to the model protein for layers. Both threonine and tryptophan levels in tropical forages seem to be well balanced, and the latter is within the desired range in half of the species analyzed.

Generally, green parts of tropical plants have higher levels of tryptophan compared to the seeds. This pattern of amino acid profiles does not need a huge amount of light, because poultry diets and plant species are frequently mixtures of numerous ingredients, which when combined, should integrate with each other to cover the nutritional requirements. Forages can have further positive effects when included in diets of monogastrics. Evidence has been provided that hens' fertility improved when their diet included 14% grass meal [16]. The inclusion of lucerne and grass meal in their diet declined the level of cholesterol in their eggs [17]. With regard to the nutritional value of feed, it has been proposed that feed influences not only the critical nutrients they contain, but also their digestibility, and hence their actual availability. Both dietary fiber and plant secondary compounds (with anti-nutritional factors or toxic) are the major factors which can strongly affect digestibility.

6. Anti-Nutritional Factors and Chemical Constraints

It is well-documented that plants consist of a variety of simple to extremely complex mixtures, many of which have been well-known and described. It seems that almost all of these constituents of plants have a defense function against abiotic and biotic stresses, and more than 1200 classes assist to protect against herbivores. The study implemented by Makkar [18] clearly showed that these compounds were not implicated in the plant primary biochemical passageways for cell reproduction and growth.

The most common major groups are alkaloids, polyphenols, saponins, cyanogenic glycosides, steroids, amino acids and toxic proteins, non-protein amino acids, phyto-hemaglutinins, oxalic acid and triterpenes [19], and were either toxic or acted as anti-nutritive factors (ANF). Anti-nutritive factors are defined as "substances generated in natural feed ingredients by the normal metabolism of (plant) species and (interacting) by different mechanisms, e.g., inactivation of some nutrients, interference with the digestive process or metabolic utilization of feed which exert effects contrary to optimum nutrition. Being an ANF is not an intrinsic characteristic of a compound but depends upon the digestive process of the ingesting animal" [20]. Consequently, plants that cause harmful effects to humans or other mammals may often be vastly toxic for fish, birds, and insects or other small animals [21]. The efficacy of leaves, pods, edible twigs of trees and shrubs as animal feed is in narrow usage because of the presence of ANFs. In general, ANFs are not harmful, but may cause toxicity during periods of shortage when animals consume large quantities of ANF-rich feed.

6.1. Polyphenolic Compounds

Polyphenols are a main group often interrelated with odor, taste, and color. Flavonoids (i.e., monomeric elements of condensed tannins), lignane, and cumarine are the major mediators. Condensed tannins (CT) are complex heat-stable phenolic constituents and are widespread in abundant plants, mainly shrub legumes such as *Gliricidia sepium* Jacq., *Leucaena leucocephala* Lam., *Acacia* species, and *Albizia falcataria* L. Proteins bind tannins side by side by H bonds and hydrophobic connections. These phenomena may be related to reduce the availability and digestibility of protein [12] and other nutrients such as fibers and starch. Another limiting factor is their astringent taste, which in many cases decreases palatability, thus the animal will not engorge on it.

6.2. Tannins

Tannins have molecular weight of more than 5 KD, and are water soluble. Tannins have the ability to bind with proteins and minerals. There are two different groups of tannins: condensed and hydrolysable tannins. Grain legumes, seeds, and forages have a wide distribution of condensed tannins. Livestock are sensitive to this type of tannin (i.e., condensed), while goats are more able to adapt to high amounts of tannins. Cattle and sheep are sensitive to condensed tannins, while goats are more resistant [22]. Tannins negatively impact digestive processes, though they may bind with endogenous enzymes such as of trypsin, amylase chemotrypsin, and lipase, and have the ability to bind with vitamin B_{12}. Additionally, it has been reported that they can cause damage to intestinal cells, and interfere with iron absorption, and tannins may possibly generate a toxic effect [23]. Tannins cause reductions in protein digestibility in humans and animals, inhibit digestive enzymes, and increase nitrogen in the feces. The influence of tannins on animal performance have been studies [24], they are recognized to be responsible for declines in growth rates, feed intake, and feed efficiency by binding with proteins and reducing protein digestibility. If tannin levels in the diet become high, microbial enzyme activities together with cellulose and intestinal digestion may be depressed.

6.3. Lectins

Lectins are glycoproteins commonly scattered in grain legumes and certain oil seeds (including soybeans) which possess an affinity for specific sugar molecules, and are categorized by their capability to link with receptors of carbohydrate membranes [25]. Lectins are categorized as growth inhibitory (*Glycine max* L., *Phaseolus lunatus* L., *Amaranthus cruentus* L., *Dolichos biflorus* L.), toxic (*Canavalia ensiformis* L., *Phaseolus vulgaris* L.) [26], or fundamentally beneficial or non-toxic (seeds of *Vigna umbellate* Thunb., *Vigna subterranean* L., and *Vigna unguiculate* L.) [27]. The efficacy of lectins either toxic or non-toxic is dependent on the developmental stage and part of the plant. The toxic consequences of lectins is that they usually coagulate the erythrocytes, and thus may depress the immune system [12] or interrupt nutrient absorption in the small intestine by shedding the brush boundary membrane of the intestinal absorptive cells [18]. Lectins have the ability to immediately connect with the epithelial mucosa of the intestines, interrelating the enterocytes [25], and interfering with the absorption and transportation of 0.01% free gossypol within some low gossypol cotton levels (mainly carbohydrates) during ingestion, causing epithelial lesions within the intestine. Some types of lectins are ordinarily reported as being unstable, because their stability differs among plants, many lectins are resistant to hydrolases by dry thermal and require the existence of humidity for whole destruction [25].

Three physiological reactions can be observed when used by extra-sensitive individuals. Firstly, they can cause nutrition deficiencies through severe intestinal damage and reduced digestion. Secondly, they can induce IgM and IgG antibodies causing food sensitiveness and other immune responses [10,25]. Finally, they can link to erythrocytes, concurrently with immune factors, causing anemia and hemagglutination. Generally, lectins can modify host resistance to infection, causing failure to thrive and can even lead to death in animals.

6.4. Phytic Acid

Phytic acid is one of main concerns for nutrition and health management in humans [28]. At physiological pH, the phytate molecule is negatively charged, and binds to nutritionally indispensable divalent cations, such as calcium, iron, magnesium, and zinc. These patterns of phytase are insoluble complexes, this mean trace elements are unobtainable for absorption [29]. The levels of phytic acid present were 0.624 to 1.0% (mean 0.862%). The suggestion has been made by Egli et al. [30] that there are higher levels of phytate in millet and pigeon pea. On the other hand, the inconsistencies may be expected, as according to Reddy et al. [31], phytate values differs in cereals and legumes depending on soil type, diversity, and cultivate type.

6.5. Saponins

Saponins can originate in numerous food plants such as *Amaranthus hypochondriacus* L., *Atriplex hortensis* L., *Chenopodium quinoa* Willd. [32], *Brachiaria brizantha* Hochst., *B. decumbens* Sm. [29], and *Medicago sativa* L. They are recognized as heat-stable, form a soapy froth when interacting with water, and can adjust cell wall permeability, leading to hemolysis and to photosensitization [29]. Saponins were deemed as toxic compounds, since they appeared to be exceedingly toxic to fish and endothermic animals, and many of them influenced strong hemolytic activity. The high levels of saponins in dietary plants may create an acidic taste and astringency. Feed intake or consumption can be limited by the harsh taste of saponin. In the last decades, saponins were documented as anti-nutrient ingredients, due to their reverse influences, such as for growth declines and decrease in food consumption due to the bitterness and throat-irritating activity of saponins. Furthermore, saponins were found to shrink the bioavailability of nutrients and inhibit digestive enzyme activities (trypsin and chymotrypsin) [25] consequently reducing protein digestibility.

6.6. Toxic Amino Acids

Non-protein amino acids are established in many plants as unconjugated forms, mainly in legumes, with the highest levels in the seeds. For instance, *Leucaena leucocephala* encloses mimosine, which binds totopyridoxalphosphate and minerals [18], leading to declines in the activity of the enzymes that involve them as co-factors, and eventually suppressing metabolic passageways. A study by Sastry and Rajendra [33] showed that the teratogenicity effects of non-protein amino acids can impede the reproductive process, leading to loss of wool and hair, and even to death.

Most legumes seeds such as the *Canavalia* species, *Medicago sativa* [34], and *Vicia ervilia* [35] consist of canavanine. The inhibition effect of canavanine on the development of insects is due to the competition for the irreplaceable amino acid arginine. Poultry are much more vulnerable to canavanine than mammals due to the antibiosis of arginine with lysine in birds. This disruption may be lead to autoimmune-like infections affecting the skin and kidneys. There are various plants which have been identified for their toxic effects such as L-DOPA, which is currently in the *Mucuna* species, and is cytotoxic [36] leading to haemolytic anaemia. Lathyrogenic amino acids, like BCNA (β-cyanoalanine), DABA (α,γ-diaminobutyric acid), ODAP(β-N-oxalyl-α,β-diaminopropionic acid), and BAPN (β-aminopropionitrile) are neurotoxic and occur in the *Lathyrus* species and *Vicia sativa* [37]. Canavanine is highly toxic and present in *Canavalia ensiformis* seeds, and is a potent insecticide [38].

6.7. More ANFs

Increasing attention has been paid to identifying other anti-nutritional factors and to stopping their negative impacts on mammals, birds, and other animals. It has been found that some proteins are heat-stable antigenics and heat-labile cyanogens proteins, amongst others. Cyanogenic glycosides, such as hydrocyanic acid, linamarin, and lotaustralin, which are commonly present in cassava (*Manihot esculenta*) and also in *Phaseolus*, *Acacia*, and *Psophocarpus*, decrease performance and cause cyanide intoxication. If, however, "the diet is adequately supplemented with proteins, particularly with sulfur-containing amino acids, and iodine", it is safe to feed to livestock [39].

7. Processes to Improve Nutritional Value of Forages

Forages and grain legumes which contain some ANFs in their meal and seeds, and which decline the availability of nutrients [40], surely have adverse influences on animal performance and human nutrition. As pointed out earlier, these ANFs are descried above such as tannins, phenols, toxic amino acids, phytic acids, lectins, trypsin inhibitors, and cyanogenic glycosides. It was found that those ANFs in low concentration [40,41] may reduce the nutritional quality of forages and grain legumes, even if they are used for feeding animals. As a result, those ANFs need to be removed or reduced, and hence the nutritional values and bioavailability of forages and grain legumes will be improved. Numerous management procedures such as germination, fermentation, thermal management (i.e., boiling, autoclaving, and cooking), soaking, and de-hulling procedures have been useful to decrease or eliminate the concentration of these compounds from forages and grain legumes [41,42]. In general, these processing techniques superficially enhance not only the palatability and flavor of legumes, but also increase the bioavailability of nutrients and protein digestibility by destroying the ANFs [43].

7.1. Heat Treatments

Heat remediation encompasses oven and sun-drying, autoclaving, roasting, and boiling, which ordinarily decreases the content of heat-labile ANFs. It has been proposed that sun drying cassava leaves (*Manihot esculenta*) reduces hydrogen cyanide from 20 mg/kg in the leaf meal, compared with 190 mg/kg in the meal of fresh leaves. Montilla et al. [44] reported that feeding laying hens with sun-dried *Gliricidia sepium* resulted in better performance than those fed with the oven-dried legumes, but the effects of type of drying are not clear on the feed quality. Thermal management significantly reduced the trypsin-inhibitory activity of seeds of *Cajanus cajan* [45,46], *Glycine max* [47], *Psophocarpus tetragonolobus* [48], and *Arachis hypogaea* [49].

In this context, haemagglutinin can be altogether discarded by roasting. Both thermal management, autoclaving or roasting seeds of *Phaseolus vulgaris* decreased its tannin content by 30–40%, and this was minimized by autoclaving [50] and de-hulling by dry heat [51], which expressively reduced the content of L-DOPA in seeds of *Mucuna pruriens* L. Recently, to increase nutritive values of guava seeds, it was shown that they must be roasted at 150 °C for 10 min [5]. Phytic acid and tannin content were significant reduced by roasting guava seeds [5], and the highest declines were affected by roasting at 150 °C for 20 min (520.1 and 61.36%, respectively). Thermal processes destroy the naturally occurring ANFs and is implemented in order to decline anti-nutrients in plant-based foods, thus enhancing the nutritive value of isolated protein [5,52].

7.2. Soaking

Several approaches have been implemented to reduce the phytic acid level in feed, especially cereal which becomes poor in quality due to such anti-nutrients. One of these approaches is to soak the grains. Vijayakumari et al. [53] revealed that treatment by soaking grains in water for 18 h decreased the phytate level of *Mucuna monosperma* by up to one-third of the original content. Soaking in water at room temperature overnight has been linked to a 36% reduction in phytic acid in kidney beans [41]. The loss in phytates during soaking of the tested samples may be due to leaching of phytate ions into the soaking water under the influence of a concentration of a gradient (difference in chemical potential), which governs the rate of diffusion. Soaking in a mixed-salt solution (0.75% citric acid + 0.5% Na_2CO_3 + 1.5% $NaHCO_3$) led to a significant reduction in anti-nutritional factors such as tannins, phytase, orthodihydroxy phenols, and phenols in pigeon pea hybrids [40].

7.3. Pelleting

Nutritive value and voluntary feed intake can be determined by feeding texture. The digestibility of protein and starch in chicks as well as apparent metabolizable energy values of *Vicia faba* [54] has been increased by feed pelleting. Pelleting processes of three herbaceous legumes *Lablab purpureus* L., *Calopogonium mucunoides* Desv., and *Mucuna pruriens* [55] reduced the anti-nutritional factors and fiber fraction contents.

7.4. Hulling

Hulling is a simple method for removing ANFs such as tannins, which are mostly focused in the seed coat [56]. De-hulling declines the tannin level from 22.0 to 5.3 mg/100 g in *Phaseolus vulgaris* seeds. This technique might be an opportunity for farmers, such as coffee growers, who have other uses for a de-hulling mill. Other opportunities for small-scale milling are explained by Jonsson et al. [57].

7.5. Germination

Germination activates endogenous enzymes, which attack most anti-nutrients and enhances the nutritional value of grains [58]. Germination triggers endogenous enzymes and improves the nutritional value of grains through an onslaught against most of anti-nutrients [58]. Nevertheless, germination can be difficult to manage as seedlings tend to share molds and are simply spoiled. Feeding of germinated seeds must occur immediately or they should be dried, otherwise the cost will rise. Germination diminishes esphytic acid, trypsin inhibitors, certain lectins, and galactosidesin *Glycine max* [59], and compared to raw seeds, improves the in vitro starch digestibility of *Vigna radiate*, *Vigna unguiculata*, and *Cicer arietinum*, similar to the enhancements found through fermentation and pressure cooking [60].

Soaking followed by germination, decreased the trypsin inhibitory activity of *Cajanus cajan* and *Phaseolus vulgaris* seeds by 26–53%, condensed tannins by 14–36%, and phytic acid by 41–53%, while the invitro protein digestibility, and thiamine and vitamin C levels were augmented significantly, in addition to an alteration in the mineral arrangement [61]. Germination of *Lupinus albus* and *Lupinus luteus* for 96 and 120 h led to peak phytase activity, respectively [58].

7.6. Fermentation

Fermentation is a process, which occurs under anaerobic conditions by microbes, which have the capacity to ferment carbohydrates into organic acids and/or alcohols. Ensiling is an appropriate fermentation process for both whole crop forage and grains. The main product produced during fermentation is lactic acid, and the diminishα-amylase and trypsin inhibitor activity and tannins in *Sphenostylis stenocarpa* seeds were reduced by up to 100%, which is dissimilar to cooking [62]. It also reduced alpha-galactosides and cyanogenic glycosides by 85%, compared with only 10–20%when cooked. The increasing in vitro protein digestibility decreased minerals and affected various vitamin profiles through fermentation of *Phaseolus vulgaris* grains and grain meal [63], reducing α-galactosides, trypsin inhibitory activity and tannin content in seed meal. Fermenting *Mucuna* to tempera in traditional Indonesian food hydrolyzes 33% of phytic acid and reduces L-DOPA by 70% [64]. Solid state fermentation of *Cicer arietinum* gives higher digestibility of protein and lysine, and decreases tannin levels up to 13% and phytic acid concentration by 10%in raw chickpea flour [65].

Good fermentation management [66], which is achievable for smallholders, is required to avoid notable losses of tryptophan and lysine [67], and can even benefit from increased lysine content. Further information on ensiling and silo types is available through the Food and Agriculture Organization of the United Nations (FAO) [4] and Heinritz et al. [68]. All over the world, sorghum is considered a main cereal, and is drought resistant. It has been reported that it is an essential and principle source of protein, energy, and minerals in diets for both animals and humankind in tropical and subtropical areas [4]. Previous studies have shown that sorghum contains anti-nutritional factors like tannin, phytic acid, cyanogenicglucoside, oxalate, and trypsin inhibitor [69,70]. For these reasons, it is categorized as having low nutritional values. Fermentation of sorghum diminished oxalate, phytate, and tannins by 49.1%, 40%, and 16.12%, respectively [70].

8. Non-Conventional Ingredients as a Protein Source

Several reports have highlighted the proximate analyses of many non-conventional protein sources which can be used in poultry production (Table 1). Generally, average daily protein consumption in rural, resource-poor areas is less than 9 g of animal protein (capita/day), compared to over 60 g (person/day) as reported by FAO recommendations for daily protein consumption [71,72]. At the family level, alternative sources of protein for smallholders in rural areas must be provided, as they areoften resource-poor areas, and it is difficult to access commercial diets in these areas.

Table 1. The proximate chemical analysis (%) of some alternative protein sources for poultry feeding.

Items	Dry Matter	Crude Protein	Crude Fiber	Ether Extract	Ca	P	Reference
Moringa leaves	80	29.7	22.5	4.38	2.78	0.26	[73]
Moringa leaves	94	27.2	40	17.1	-	-	[74]
Leuceana leucocephala leaves	88	25.9	40	-	2.36	0.23	[75]
Neem (*Azadirachta indica*) leaves	92	20.68	16.6	4.13	-	-	[76]
Cassava (*Manihot escsulenta*) leaves	95.5	26.3	19.7	7.3	-	-	[76]
Gliricidia sepium	89.3	22.9	17.15	8.8	-	-	[77]
Pawpaw (*Carica papaya*) leaves	93.2	26.3	14.8	-	3.2	-	[78]
Cabbage leaves	87.9	14	35	2.0	3.2	-	[79]
Ipomea batata leaves meal	88.7	25.66	12.76	3.06	-	-	[80]

As cited before, in the developing and developed economies of the world, non-ruminant animal production, especially the poultry industry, plays an essential role as a principal source of animal protein including both meat and eggs. The feed cost in poultry production represents around 70%of the total cost of production. Therefore, the profitability of this industry depends largely on the quality and economics of feed production [81,82]. Expansion of the industry will therefore mainly be determined by the sufficient availability and affordability of good quality feed for birds and subsequently good poultry products, such as eggs and meats for consumers [83]. For intensive enterprise, inadequacies in nutrient supply often leads to a drop in egg production, as well as declines in growth performance on the part of broilers for meat production.

9. Application of Alternative Resources of Protein for Chickens

With increasing demand for animal protein and the development of poultry industries in all regions of the world, especially regions with extreme poverty, it is necessary to save alternative protein resources for smaller breeds of chicken. Different sources of alternative feed available for feeding chickens will raise the standard of living for small breeders. Current animal industries faceconflicting requests to produce large volumes of high-quality food at low prices. Now, nutritional elucidations have come to be even more imperative to disband such instances, and this can be accomplished by taking full advantage of the alternative feed resources, such as tropical plants, in poultry diets. Furthermore, one of the ways of reducing the cost of animal production in developing countries, and therefore making protein available to people at cheaper prices, is by using agricultural by-products and tropical plants, which are not directly used by humans as food to feed livestock [84]. As with other vegetables and fruits, great amounts of waste are produced through packaging, harvest, and processing. It has been estimated that about 30–50% of total production is discarded as waste, which consists mostly of leaves or seeds [85].

As waste production increases, there is an affiliated increase in the quantity of residues produced. These wastes or residues are often leached into the environment where they cause major negative environmental impacts (e.g., nitrate percolating into water sources). The inclusion of diets with 2.5 g/kg neem (*Azadirachta indica*) leaf meal (AILM) to broiler chicken diets improved blood indices without any deleterious effects when compared to birds fed the control diet [86]. In addition, they suggest that the dietary supplementation of *Azadirachta indica* leaf meal may lead to the expansion of low-cholesterol chicken meat. Olabode and Okelola [87] concluded that supplementing neem leaf meal (AILM) to laying birds by up to 8% did not have any negative influences on egg quality, serum biochemical indices, and the final consumer product.

Compared to conventional commercial feeds, *Moringa oleifera* meal (MOLM) as a protein supplement in broiler diets at the 25% inclusion level showed similar results in weight and growth rates [88]. Pawpaw (*Carica papaya*) leaf meal (CPLM) was reportedly incorporated at the 2% level in the diet of finishing broilers. A significant improvement of 14% in growth performance was observed compared to the birds on the control diet, and the carcasses and organoleptic indices of the birds were equally recorded with positive corresponding economic returns as observed by the significantly lower feed cost/kg gain [89]. Ayssiwede et al. [90] found that the inclusion of *Leucaena leucocephala* leaves meal (LLLM) in the diet at the 21% level had no significant adverse effect on feed intake, average daily weight gain, feed conversion ratio, and nutrients utilization (except ether extract) of adult indigenous Senegal chickens, but significantly ($p < 0.05$) improved the crude protein and metabolizable energy utilization in birds fed at the 7% inclusion level. This improvement may be attributed to the nutritional quality of protein of LLLM, and was probably due to their higher percentage of sulfur amino acids. As previously noted, because of the high price of protein ingredients such as fish meal, and soybean meal, using LLLM as an alternative protein source for feeding chickens could improve small holders poultry nutrition and productivity [90], and allow for alow cost of production and enhancement of their incomes.

On the other hand, at a 5% inclusion level, cassava (*Manihotesculenta*) leafmeal (MELM) in broiler finisher diets was reported to confer a significant ($p < 0.05$) increase in feed intake, body weight gain, feed conversion ratio, and organ weight of birds without any deleterious effects [91] over those with 10 and 15% inclusion levels.

Kagya-Agyemang et al. [92] recommended an inclusion level of not more than 5% *Gliricidia sepium* leaf meal (GSLM) in broiler diets as he recorded a better carcass dressing percentage at this level, while a progressive decrease in carcass dressing percentage was observed at higher inclusion rates, with a 15% inclusion level having a significantly ($p < 0.05$) lower carcass dressing percentage. However, there was a corresponding increase in the intensity of yellow pigmentation of the skin, shanks, feet, and beaks of the birds. This impairment might be due to increasing fiber content and anti-nutritional factors present in Gliricidia leaf meal (GLM). These constituent include alkaloids, tannins, and nitrates that can decrease the palatability of diets with GSLM [92].

Cabbage leaves are categorized by high crude protein (17%) and mineral levels, especially Ca, S, and Mn. However, some of these wastes or residuals might contain anti-nutritional factors, which may reduce animal growth performance [93].

As mentioned above, several anti-nutritional factors have been described as treatments for the alleviation of this situation. Mustafa and Baurhoo [79] evaluated the replacement of soybean meal with dried cabbage leaf residue (DCR) on broiler growth performance and nutrient digestibility. The results concluded that inclusion of dietary DCR up to 9% of the diet had no adverse influences on broiler performance. Sugar beet (*Beta vulgaris*) is a rich source of protein, carbohydrates, and has high levels of essential vitamins and micro and macro elements. Significant increases of live weight in growing geese was shown with 5–10% sugar beet pulp.

10. Conclusions

Smallholders in poultry production systems have some constraints in obtaining commercial feed. However, in tropical and subtropical areas there are large numbers of diverse forage species as protein options for their poultry, which are locally available. There is a multiplicity of options related to agricultural suitability and nutrients, and yields contents, and nutritional obstacles may be relatively or abundantly overcome by suitable processing techniques. At the farm level, successes of worthy economic returns can be achieved through ecological conditions such as individual decisions, technical and labor requirements, and already-available feed materials. Knowledge of the characteristics of nutrition, agronomics, and secondary metabolites compounds of forage species allows small breeders of poultry to make better choices for feeding and raising their poultry in rural, resource-poor areas. Creative approaches are required to appropriate forage-based feed solutions for poultry into present smallholder systems and further systematic investigation is required to describe the actual value of some less-widespread forage species for various animal species.

Author Contributions: Conceptualization, M.E.A.E.-H.; Writing-Original Draft Preparation, S.A.A. and M.R.; Writing-Review & Editing, M.E.A.E.-H.; Visualization, S.A.A.; Administration, M.E.A.E.-H.

Funding: This research received no external funding.

Acknowledgments: Authors thank their respective universities for their support.

Conflicts of Interest: The authors declare no conflict of interest.

References

1. OECD and Food and Agriculture Organization of the United Nations Meat. *OECD-FAO Agricultural Outlook 2010*; OECD Publishing: Paris, France, 2010; pp. 147–158.
2. McDermott, J.J.; Staal, S.J.; Freeman, H.A.; Herrero, M.; Van de Steeg, J.A. Sustaining intensification of smallholder livestock systems in the tropics. *Livest. Sci.* **2010**, *130*, 95–109. [CrossRef]
3. FAOSTAT. Production Quantity of Soybeans 2016. Available online: http://faostat.fao.org/site/567/DesktopDefault.aspx?PageID=567#ancor (accessed on 10 March 2018).
4. Food and Agricultural Organization (FAO). *Official Agricultural Statistics*, 18th ed.; Food and Agricultural Organization: Quebec, QC, Canada, 2006; pp. 1002–1004.
5. El Anany, A.M. Nutritional composition, antinutritional factors, bioactive compounds and antioxidant activity of guava seeds (Psidium Myrtaceae) as affected by roasting processes. *J. Food Sci. Technol.* **2013**. [CrossRef] [PubMed]
6. National Academy of Sciences (NAS). *Tropical Legumes: Resources for the Future*; National Academy of Sciences (NAS): Washington, DC, USA, 1979.
7. Karaimu, P. ILRI Gene Bank Manager Elected 'Fellow' of the Prestigious Society of Biology. 2011. Available online: http://www.ilri.org/ilrinews/index.php/archives/tag/ilri-genebank (accessed on 10 March 2018).
8. CIAT. Forages Collection. Centro Internacional de Agricultura Tropical. 2011. Available online: http://isa.ciat.cgiar.org/urg/foragecollection.do (accessed on 10 March 2018).
9. Joze-fiak, D.; Rutkowski, A.; Martin, S.A. Carbohydrate fermentation in the avian ceca: A review. *Anim. Feed Sci. Technol.* **2004**, *113*, 1–15. [CrossRef]
10. Fereidoon, S. *Antinutrients and Phytochemicals in Food*; Developed from a Symposium Sponsored by the Division of Agricultural and Food Chemistry; ACS Symposium Series; American Chemical Society: Washington, DC, USA, 2012. [CrossRef]
11. Shayo, C.M.; Uden, P. Nutritional uniformity of crude protein fractions in some tropical browse plants estimated by two in vitro methods. *Anim. Feed Sci. Technol.* **1999**, *78*, 141–151. [CrossRef]
12. Jeroch, H.; Drochner, W.; Simon, O. *Ernährung Landwirtschaftlicher Nutztiere*; UTB: Stuttgart, Germany, 1999.
13. Campbell, G.L.; van der Poel, A.F.B. Use of enzymes and process technology to inactivate antin-utritional factors in legume seeds and rapeseed. In *Recent Advances of Research in Antinutritional Factors in Legume Seeds and Rapeseed, Proceedings of the Third International Workshop, Wageningen, The Netherlands, 8–10 July 1998*; Jansman, A.J.M., Huisman, J., van der Poel, A.F.B., Eds.; Wageningen University Research: Wageningen, The Netherlands, 1998; pp. 377–386.
14. Schultze-Kraft, R.; Peters, M. Tropical legumes in agricultural production and resource management: An overview. *Giessener Beitraege zur Entwicklungsforschung* **1997**, *24*, 1–17.
15. Savon, L. Tropical roughages and their effect on the digestive physiology of monogastric species. *Cuban J. Agric. Sci.* **2005**, *39*, 475–487.
16. Davtyan, A.; Manukyan, V. Effect of grass meal on fertility of hens. *Ptisevodstvo* **1987**, *6*, 28–29.
17. Rybina, E.A.; Reshetova, T.A. Digestibility of nutrients and biochemical values of eggs in relation to the amount of lucerne and grass meal and the quality of supplementary fat in the diet of laying hens. *Trudy Uzbekskogo Nauchno Issledovatel'skogo Instituta Zhivotnovodstva* **1981**, *35*, 148–152.

18. Makkar, H.P.S. Antinutritional factors in animal feedstuffs-mode of actions. *Int. J. Anim. Sci.* **1991**, *6*, 88–94.
19. Liener, I.E. *Toxic Constituents of the Plants Food-Stuffs*; Academic Press: London, UK, 1980.
20. Aganga, A.A.; Tshwenyane, S.O. Feeding values and anti-nutritive factors of forage tree legumes. *Pak. J. Nutr.* **2003**, *2*, 170–177.
21. Kumar, R.; D'Mello, J.P.F. Anti-nutritional factors in forage legumes. In *Tropical Legumes in Animal Nutrition*; D'Mello, J.P.F., Devendra, C., Eds.; CABI Publishing: Wallingford, UK, 1995; pp. 95–133.
22. Kumar, R.; Vaithiyanathan, S. Occurrence, nutritional significance and effect on animal productivity of tannins in tree leaves. *Anim. Feed Sci. Technol.* **1990**, *30*, 21–38. [CrossRef]
23. Butler, L.G. Effects of condensed tannins on animal nutrition. In *Chemistry and Significance of Condensed Tannins*; Hemingway, R.W., Karchesy, J.J., Eds.; Plenum Press: New York, NY, USA, 1989; pp. 391–402.
24. Aletor, V.A. *Anti-Nutritional Factors as Nature's Paradox in Food and Nutrition Securities*; Inaugural Lecture Series 15; The Federal University of Technology, Akure (FUTA): Akure, Nigeria, 2005.
25. Gerpacio, A.L.; Princesa, A.O. Effects of heat treatment and fat extraction on the nutritive value of winged bean seed meal for broilers (Philippines). *Anim. Prod. Technol.* **1985**, *1*, 33–34.
26. Grant, G. Anti-nutritional effects of dietary lectins. *Asp. Appl. Biol.* **1989**, *19*, 51–74.
27. Grant, G.; More, L.J.; McKenzie, N.H.; Dorward, P.M.; Buchan, W.C.; Telek, L.; Pusztai, A. Nutritional and haemagglutination properties of several tropical seeds. *J. Agric. Sci.* **1995**, *124*, 437–445. [CrossRef]
28. Kumar, V.; Sinha, A.K.; Makkar, H.P.S.; Becker, K. Dietary roles of phytate and phytase in human nutrition: A review. *Food Chem.* **2010**, *120*, 945–959. [CrossRef]
29. Brum, K.B.; Haraguchi, M.; Garutti, M.B.; No-brega, F.N.; Rosa, B.; Fioravanti, M.C.S. Steroidal saponin concentrations in *Brachiariadecumbens* and *B. brizantha* at different developmental stages. *Ciencia Rural* **2009**, *39*, 279–281. [CrossRef]
30. Egli, I.; Davidsson, L.; Zeder, C.; Walczyk, T.; Hurrell, R. Dephytinization of a complementary food based on wheat and soy increases zinc, but not copper, apparent absorption in adults. *J. Nutr.* **2004**, *134*, 1077–1080. [CrossRef] [PubMed]
31. Reddy, N.R.; Sathe, S.K.; Salunkhe, D.K. Phytates in legumes and cereals. In *Advances in Food Research*; Academic Press: Cambridge, MA, USA, 1982; Volume 28, pp. 1–92.
32. Cheeke, P.R.; Carlsson, R. Evaluation of several crops as sources of leaf meal: Composition, effect of drying procedure, and rat growth response. *Nutr. Rep. Int.* **1978**, *18*, 465–473.
33. Sastry, M.S.; Rajendra, S. Toxic effects of subabul (*Leucaena leucocephala*) on the thyroid and reproduction of female goats. *Indian J. Anim. Sci.* **2008**, *78*, 251–253.
34. Nunn, P.B.; Bell, E.A.; Watson, A.A.; Nash, R.J. Toxicity of non-protein amino acids to humans and domestic animals. *Nat. Prod. Commun.* **2010**, *5*, 485–504. [PubMed]
35. Sadeghi, G.H.; Pourreza, J.; Samei, A.; Rahmani, H. Chemical composition and some anti-nutrient content of raw and processed bitter vetch (*Viciaervilia*) seed for use as feeding stuff in poultry diet. *Trop. Anim. Health Prod.* **2009**, *41*, 85–93. [CrossRef] [PubMed]
36. Lee, J.; Kim, Y.; Park, S.; Lee, M. Effects of tributyltin chloride on L-DOPA-induced cytotoxicity in PC12 cells. *Arch. Pharm. Res.* **2006**, *29*, 645–650. [CrossRef] [PubMed]
37. D'Mello, J.P.F.; Walker, A.G. Detoxification of Jack Beans (*Canavaliaensiformis*)—Studies with young chicks. *Anim. Feed Sci. Technol.* **1991**, *33*, 117–127. [CrossRef]
38. Rosenthal, G.A. L-Canavanine and L-Canaline: Protective Allelochemicals of Certain Leguminous Plants. In *Plant Resistance to Insects, Proceedings of the Symposium Held at the 183rd Meeting of the American Chemical Society, Las Vegas, NV, USA, 28 March–2 April 1982*; Hedin, P.A., Ed.; American Chemical Society: Washington, DC, USA, 1983; pp. 279–290.
39. Tewe, O. Indices of cassava safety for livestock feeding. In Proceedings of the International Workshop on Cassava Safety, Ibadan, Nigeria, 1–4 March 1994; Volume 375, pp. 241–250.
40. Devi, R.; Chaudhary, C.; Jain, V.; Saxena, A.K.; Chawla, S. Effect of soaking on anti-nutritional factors in the sun-dried seeds of hybrid pigeon pea to enhance their nutrients bioavailability. *J. Pharmacogn. Phytochem.* **2018**, *7*, 675–680.
41. Abd El-Hady, E.A.; Habiba, R.A. Effect of soaking and extrusion conditions on antinutrients and protein digestibility of legume seeds. *Lebensm. Wiss. Technol.* **2003**, *36*, 285–293. [CrossRef]

42. Martin-Cabrejas, M.A.; Vidal, A.; Sanfiz, B.; Molla, E.; Esteban, R.; Lopez-Andreu, F.J. Effect of fermentation and autoclaving on dietary fiber fractions and antinutritional factors of beans (*Phaselous vulgaris* L.). *J. Agric. Food Chem.* **2004**, *52*, 261–266. [CrossRef] [PubMed]

43. Khandelwal, S.; Udipi, S.A.; Ghugre, P. Polyphenols and tannins in Indian pulses: Effect of soaking, germination and pressure cooking. *Food Res. Int.* **2010**, *43*, 526–530. [CrossRef]

44. Montilla, J.J.; Reveron, A.; Schmidt, B.; Wiedenhofer, H.; Castillo, P.P. Leaf meal of mouse-tail (*Gliricidiasepium*) in rations for laying hens. *Agron. Trop.* **1974**, *24*, 505–511.

45. Arif, M.; Rehman, A.; Saeed, M.; Abd El-Hack, M.E.; Alagawany, M.; Abbas, H.; Arian, M.A.; Fazlani, S.A.; Abbasi, I.H.; Ayaşan, T. Effect of different processing methods of pigeon pea (*Cajanuscajan*) on growth performance, carcass traits, and blood biochemical and hematological parameters of broiler chickens. *Turk. J. Vet. Anim. Sci.* **2017**, *41*, 38–45. [CrossRef]

46. Abd El-Hack, M.E.; Swelum, A.A.; Abdel-Latif, M.A.; Toro, D.M.; Arif, M. Pigeon Pea (*Cajanuscajan*) as an alternative protein source in broiler feed. *World's Poult. Sci. J.* **2018**, in press.

47. Liener, I.E. Implications of Antinutritional Components in Soybean Foods. *Crit. Rev. Food Sci. Nutr.* **1994**, *34*, 31–67. [CrossRef] [PubMed]

48. Igene, F.U.; Oboh, S.O.; Aletor, V.A. Nutrient and anti-nutrient components of raw and processed winged bean seeds (*Psophocarpus tetragonolobus*). *Indian J. Anim. Sci.* **2006**, *76*, 476–479.

49. Hira, C.K.; Chopra, N. Effects of roasting on protein quality of chickpea (*Cicer arietinum*) and peanut (*Arachishypogaea*). *J. Food Sci. Technol.-Mysore* **1995**, *32*, 501–503.

50. Vijayakumari, K.; Siddhuraju, P.; Janardhanan, K. Effect of different post-harvest treatments on antinutritional factors in seeds of the tribal pulse, *Mucuna pruriens* (L.) DC. *Int. J. Food Sci. Nutr.* **1996**, *47*, 263–272. [CrossRef] [PubMed]

51. Siddhuraju, P.; Vijayakumari, K.; Janardhanan, K. Chemical composition and protein quality of the little-known legume, velvet bean (*Mucuna pruriens* (L.) DC). *J. Agric. Food Chem.* **1996**, *44*, 2636–2641. [CrossRef]

52. Seena, S.; Sridhar, K.R.; Arunb, A.B.; Young, C.C. Effect of roasting and pressure-cooking on nutritional and protein quality of seeds of mangrove legume *Canavalia cathartica* from southwest coast of India. *J. Food Compos. Anal.* **2006**, *19*, 284–293. [CrossRef]

53. Vijayakumari, K.; Siddhuraju, P.; Janardhanan, K. Effect of soaking, cooking and autoclaving on phytic acid and oligosaccharide contents of the tribal pulse, *Mucuna monosperma* DC ex Wight. *Food Chem.* **1996**, *55*, 173–177. [CrossRef]

54. Lacassagne, L.; Francesch, M.; Carre, B.; Mel-cion, J.P. Utilization of Tannin-Containing and Tannin-Free Faba Beans (*Viciafaba*) by Young Chicks—Effects of Pelleting Feeds on Energy, Protein and Starch Digestibility. *Anim. Feed Sci. Technol.* **1988**, *20*, 59–68. [CrossRef]

55. Oyaniran, D.K.; Ojo, V.O.; Aderinboye, R.Y.; Bakare, B.A.; Olanite, J.A. Effect of pelleting on nutritive quality of forage legumes. *Livest. Res. Rural Dev.* **2018**, *30*, in press.

56. Vadivel, V.; Janardhanan, K. Nutritional and antinutritional characteristics of seven South Indian wild legumes. *Plant Foods Hum. Nutr.* **2005**, *60*, 69–75. [CrossRef] [PubMed]

57. Jonsson, L.O.; Dendy, D.A.V.; Wellings, K.; Bokalders, V. *Small-Scale Milling: A Guide for Development Workers*; Intermediate Technology Publications Ltd. (ITP): London, UK, 1994.

58. Muzquiz, M.; Pedrosa, M.M.; Cuadrado, C.; Ayet, G.; Burbano, C.; Brenes, A. Variation of alka-loids, alkaloids esters, phytic acid and phytaseactivity in germinated seeds of *Lupinusalbus* and *L. luteus*. In *Recent Advances of Research in Antinutritional Factors in Legume Seeds and Rapeseed, Proceedings of the Third International Workshop, Wageningen, The Netherlands, 8–10 July 1998*; Jansman, A.J.M., Huisman, J., van der Poel, A.F.B., Eds.; Wageningen University Research: Wageningen, The Netherlands, 1998; pp. 387–390.

59. Bau, H.M.; Villaume, C.; Nicolas, J.P.; Mejean, L. Effect of germination on chemical composition, biochemical constituents and antinutritional factors of soya bean (*Glycine max*) seeds. *J. Sci. Food Agric.* **1997**, *73*, 1–9. [CrossRef]

60. Urooj, A.; Puttaraj, S. Effect of processing on starch digestibility in some legumes—An in-vitro study. *Mol. Nutr. Food Res.* **1994**, *38*, 38–46. [CrossRef]

61. Sangronis, E.; Machado, C.J. Influence of germination on the nutritional quality of *Phaseolus vulgaris* and *Cajanuscajan*. *LWT-Food Sci. Technol.* **2007**, *40*, 116–120. [CrossRef]

62. Azeke, M.A.; Fretzdorff, B.; Buening-Pfaue, H.; Holzapfel, W.; Betsche, T. Nutritional value of African yam bean (*Sphenostylisstenocarpa* L): Improvement by lactic acid fermentation. *J. Sci. Food Agric.* **2005**, *85*, 963–970. [CrossRef]

63. Granito, M.; Frias, J.; Doblado, R.; Guerra, M.; Champ, M.; Vidal-Valverde, C. Nutritional improvement of beans (*Phaseolus vulgaris*) by natural fermentation. *Eur. Food Res. Technol.* **2002**, *214*, 226–231. [CrossRef]

64. Higasa, S.; Negishi, Y.; Adoyagi, Y.; Sugahara, T. Changes in free amino acids of temperature during preparation with velvet beans (*Mucunapruriens*). *J. Jpn. Soc. Food Sci. Technol.* **1996**, *43*, 188–193. [CrossRef]

65. Reyes-Moreno, C.; Cuevas-Rodriguez, E.O.; Milan-Carrillo, J.; Cardenas-Valenzuela, O.G.; Barron-Hoyos, J. Solid state fermentation process for producing chickpea (*Cicer arietinum* L.) tempeh flour. Physicochemical and nutritional characteristics of the product. *J. Sci. Food Agric.* **2004**, *84*, 271–278. [CrossRef]

66. Marcinakova, M.; Laukova, A.; Simonova, M.; Strompfova, V.; Korenekova, B.; Nad, P. A new probiotic and bacteriocin-producing strain of *Enterococcus faecium* EF9295 and its use in grass ensiling. *Czech J. Anim. Sci.* **2008**, *53*, 336–345. [CrossRef]

67. Blandino, A.; Al-Aseeri, M.E.; Pandiella, S.S.; Cantero, D.; Webb, C. Cereal-based fermented foods and beverages. *Food Res. Int.* **2003**, *36*, 527–543. [CrossRef]

68. Heinritz, S.; Martens, S.D.; Avila, P.; Hoedtke, S. The effect of inoculant and sucrose addition on the silage quality of tropical forage legumes with varying ensilability. *Anim. Feed Sci. Technol.* **2012**, *174*, 201–210. [CrossRef]

69. Etuk, E.B.; Okeudo, N.J.; Esonu, B.O.; Udedibie, A.B.I. Antinutritional factors in sorghum: Chemistry, mode of action and effects on livestock and poultry. *Online J. Anim. Feed Res.* **2012**, *2*, 113–119.

70. Ojha, P.; Adhikari, R.; Karki, R.; Mishra, A.; Subedi, U.; Karki, T.B. Malting and fermentation effects on antinutritional components and functional characteristics of sorghum flour. *Food Sci. Nutr.* **2018**, *6*, 47–53. [CrossRef] [PubMed]

71. Afolayan, M.O.; Afolayan, M. Nigeria oriented poultry feed formulation software requirements. *J. Appl. Sci. Res.* **2008**, *4*, 1596–1602.

72. Boland, M.J.; Rae, A.N.; Vereijken, J.M.; Meuwissen, M.P.M.; Fischer, A.R.H.; Van Boekel, M.A.J.S.; Rutherfurd, S.M.; Gruppen, H.; Moughan, P.J.; Hendriks, W.H. The future supply of animal-derived protein for human consumption. *Trends Food Sci. Technol.* **2013**, *29*, 62–73. [CrossRef]

73. Sarwatt, S.V.; Milangha, M.S.; Lekule, F.P.; Madalla, N. *Moringa oleifera* and cottonseed cake as supplements for small holder dairy cow fed Napier grass. *Livest. Res. Rural Dev.* **2004**, *16*, 12–18.

74. Yameogo, C.W.; Bengaly, M.D.; Savadogo, A.; Nikiema, P.A.; Traore, S.A. Determination of chemical composition and nutritional values *Moringa oleifera* leaves. *Pakist. J. Nutr.* **2011**, *10*, 264–268. [CrossRef]

75. Ojo, O.A.; Fagade, O.E. Persistence of Rhizobium inoculants originating from *Leucaena leucocephala* fallowed plots in Southwest Nigeria. *Afr. J. Biotechnol.* **2002**, *1*, 23–27.

76. Esonu, B.O.; Opara, M.N.; Okoli, I.C.; Obikaonu, H.O.; Udedibie, C.; Iheshiulor, O.O.M. Physiological responses of laying birds to Neem (*Azadirachta Indica*) leaf meal based diets, body weight, organ characteristics and hematology. *Online J. Health Allied Sci.* **2006**, *2*, 4.

77. Nnaji, J.C.; Okoye, F.C.; Omeje, V.O. Screening of leaf meals as feed supplements in the culture of *Oreochromis niloticus*. *Afr. J. Food Agric. Nutr. Dev.* **2010**, *10*, 2112–2123. [CrossRef]

78. Obwanga, O.B. The Efficacy of Selected Plant Materials in Formulated Diets for Nile Tilapia, Oreochromis niloticus (L.). Master's Thesis, Egerton University, Nakuru, Kenya, 2010.

79. Mustafa, A.F.; Baurhoo, B. Evaluation of dried vegetables residues for poultry: II. Effects of feeding cabbage leaf residues on broiler performance, ileal digestibility and total tract nutrient digestibility. *Poult. Sci.* **2017**, *96*, 681–686. [CrossRef] [PubMed]

80. Sun, H.; Mu, T.; Xi, L.; Zhang, M.; Chen, J. Sweet potato (*Ipomoea batatas* L.) leaves as nutritional and functional foods. *Food Chem.* **2014**, *156*, 380–389. [PubMed]

81. Food and Agricultural Organization/World Health Organization/United Nations University. *Protein and Amino Acid Requirements in Human Nutrition*; Report of Joint FAO/WHO/UNU Expert Consultation; World Health Organization Technical Report Series # 935; World Health Organization: Geneva, Switzerland, 2007.

82. Banson, K.E.; Muthusamy, G.; Kondo, E. The import substituted poultry industry; Evidence from Ghana. *Int. J. Agric. For.* **2015**, *5*, 166–175.

83. Adejimi, O.O.; Hamzat, R.A.; Raji, A.M.; Owosibo, A.O. Performance, nutrient digestibility and carcass characteristics of Broilers fed cocoa pod husks-based diets. *Niger. J. Anim. Sci.* **2011**, *13*, 61–68.

84. Asar, M.A.; Osman, M.; Yakout, H.M.; Safoat, A.M. Utilization of corn-cob meal and faba bean straw in growing rabbits diets and their effects on performance, digestibility and economical efficiency. *Egypt. Poult. Sci.* **2010**, *30*, 415–442.

85. Choi, M.H.; Park, Y.H. Production of yeast biomass using waste Chinese cabbage. *Biomass Bioenergy* **2003**, *25*, 221–226. [CrossRef]

86. Ansari, J.; Khan, S.H.; Haq, A.; Yousaf, M. Effect of the levels of *Azadirachta indica* dried leaf meal as phytogenic feed additive on the growth performance and haemato-biochemical parameters in broiler chicks. *J. Appl. Anim. Res.* **2012**, *40*, 336–345. [CrossRef]

87. Olabode, A.D.; Okelola, O.E. Effect of neem leaf meal (*Azadirachta indica* A juss) on the internal egg quality and serum biochemical indices of laying birds. (A case study at Federal College of Agriculture, Ishiagu, Ebonyi state). *Glob. J. Biol. Agric. Health Sci.* **2014**, *3*, 25–27.

88. Gadzirayi, C.T.; Masamha, B.; Mupangwa, J.F.; Washaya, S. Performance of broiler chickens fed on mature moringaoleifera leaf meal as a protein supplement to soyabean meal. *Int. J. Poult. Sci.* **2012**, *11*, 5–10. [CrossRef]

89. Onyimonyi, A.E.; Onu, E. An assessment of pawpaw leaf meal as protein ingredient for fiishing broiler. *Int. J. Poult. Sci.* **2009**, *8*, 995–998.

90. Ayssiwede, S.B.; Dieng, A.; Chrysostome, C.; Ossebi, W.; Hornick, J.L.; Missohou, A. Digestibility and metabolic utilization and nutritional value of *Leucaena leucocephala* (Lam.) leaves meal incorporated in the diets of indigenous senegal chickens. *Int. J. Poult. Sci.* **2010**, *9*, 767–776. [CrossRef]

91. Iheukwumere, F.C.; Ndubuisi, E.C.; Mazi, E.A.; Onyekwere, M.U. Performance, nutrient utilization and organ characteristics of broilers fed cassava leaf meal *(Manihot esculenta Crantz)*. *Pakist. J. Nutr.* **2008**, *7*, 13–16. [CrossRef]

92. Kagya-Agyemang, J.K.; Takyi-Boampong, G.; Adjei, M.; Karikari Bonsu, F.R. A note on the effect of *Gliricidia sepium* leaf meal on the growth performance and carcass characteristics of broiler chickens. *J. Anim. Feed Sci.* **2007**, *16*, 104–108. [CrossRef]

93. Pereira, O.; Rosa, E.; Pires, M.A.; Fernandes, F. Brassica by-products in diets trout (*Oncorhynchus mykiss*) and their effects on performance, body composition, thyroid status and liver histology. *Anim. Feed Sci. Technol.* **2002**, *101*, 171–182. [CrossRef]

agriculture

MDPI

Article

Trifolium mutabile as New Species of Annual Legume for Mediterranean Climate Zone: First Evidences on Forage Biomass, Nitrogen Fixation and Nutritional Characteristics of Different Accessions

Mariano Fracchiolla [1], Cesare Lasorella [1], Vito Laudadio [2] and Eugenio Cazzato [1,*]

[1]   Department of Agricultural and Environmental Science, University of Bari, Aldo Moro, 70125 Bari, Italy; mariano.fracchiolla@uniba.it (M.F.); cesare.lasorella@uniba.it (C.L.)

[2]   Department of DETO, Section of Veterinary Science and Animal Production, University of Bari 'Aldo Moro', Valenzano, 70010 Bari, Italy; vito.laudadio@uniba.it

*   Correspondence: eugenio.cazzato@uniba.it; Tel.: +39-080-544-2973

Received: 14 June 2018; Accepted: 4 July 2018; Published: 9 July 2018

Abstract: The present study evaluated the forage production, nitrogen fixation and the qualitative characteristics of different accessions of *Trifolium mutabile*, a new species of annual clover, collected in southern Italy. Forage traits were assessed by harvesting plants at the vegetative stage (stem elongation) and the subsequent regrowth at the flowering stage (inflorescence emergence-main shoot). From results, significant differences were found among the accessions of *T. mutabile* in terms of forage biomass production (from 5.1 to 8.2 t ha^{-1} dry matter), capacity of nitrogen fixation (58.2–76.8% Ndfa) and forage nutritional characteristics. Besides the high forage yield, the investigated accessions showed favourable values of production and quality, representing also worthy germplasm for selection programs as well as the application for possible plant cultivar registration. Moreover, it is interesting to underline that *T. mutabile* may represent a valuable alternative to commonly cultivated annual clover species due to its prolonged vegetative cycle. However, further investigations are needed to assess the self-reseeding capacity of *T. mutabile* when utilized as pasture species.

Keywords: annual clover; N fixation; biomass production; forage

1. Introduction

Annual clovers are the basis of production systems in the Mediterranean area. Historically, the most widespread clovers are berseem (*Trifolium alexandrinum* L.), crimson clover (*Trifolium incarnatum* L.), Persian clover (*Trifolium resupinatum* L.), squarrosum clover (*Trifolium squarrosum* L.), arrowleaf clover (*Trifolium vesiculosum* L.) and subterranean clover (*Trifolium subterraneum* L.) in many countries [1–3]. *T. mutabile* (Turkish clover) is an endemic species of the Mediterranean area [4], which was described for Dalmatia. In the wild it is widespread in the western Balkans (Dalmatia, Greece, Albania) and in Southern Italy, from Campania to Sicily [5], where it is represented by *T. mutabile* var. *gussoneanum*, endemic to Southern Italy. The same species has been previously evaluated by Hoveland and Alison [6], defining it as Turkish clover having dense, compact, leafy clover with leaves resembling arrowleaf clover, and being tolerant to grazing and having also better natural reseeding than arrowleaf clover.

The vegetative parts of *T. vesiculosum* do not differ from those of *T. mutabile* in the case of vigorous plants and when grown in fertile soil. The latter species, more thermophilous and xerophilous, often appears with plants of very small size [3].

It has been well documented that the forage quality of clover is a function of the composition of the individual plant parts: each differing in quality and quantity as plants mature [7–9].

Cutting is the main agronomic factor that affects morphology and the expression of yield potential, and indirectly determines nutritive value. Optimizing the use of forages in livestock feeding requires an understanding not only of its dry matter production and plant composition but also of the changes in its nutritive value. Consequently, alternative grasslands are required and the *T. mutabile*, recently found in Italy, could be an interesting option for integration into productive systems in the Mediterranean climate zone.

Therefore, the present study described the main agronomic characteristics of different accessions of *T. mutabile* collected in south of Italy and evaluated the forage production, nitrogen fixation and the qualitative characteristics in two different phenological stages.

2. Materials and Methods

2.1. Experimental Field Trial

A field trial was carried out during the years 2015–2016 in Southern Italy at Gravina in Puglia, Bari (40°53' N; 16°24' E; 415 m above sea level), on a loamy soil, characterized as sub-alkaline, medium in total N (nitrogen) (1.22‰ Kjeldahl method), high in available P (phosphorus) (74 mg kg^{-1}; Olsen method), and exchangeable K (potassium) (362 mg kg^{-1}; BaCl$_2$; TEA method). The experimental site was characterized by a summer-dry climate with a total annual rainfall of 560 mm distributed from Autumn to Spring and a mean temperature of 16 °C. During the experimental period (from November 2015 to June 2016), the total rainfall was 336 mm, and temperatures did not show any significant variation from the average.

Seedbed preparation consisted of plowing at 40 cm during summer and dish-harrowing twice during autumn. Before sowing a total of 100 kg/ha of P$_2$O$_5$ as superphosphate was used. The previous crop was wheat (*Triticum durum*). Six different accessions of *T. mutabile* collected in Apulia Region of Southern Italy (Table 1) were seeded in 18 November 2015 at a seeding rate of 10 kg/ha seed and a row spacing of 20 cm, and ryegrass (*Lolium multiflorum* cv Trinova) was used as non-fixing reference crop at seeding rate of 40 kg/ha with a 20 cm row spacing.

Table 1. Collection sites of the different *T. mutabile* accessions.

Accession Code	Collection Site	Coordinates
TMU01	Restinco	40.6422 N; 17.8657 E
TMU02	San Donaci	40.5462 N; 17.8747 E
TMU03	San Pietro Vernotico	40.5353 N; 18.0590 E
TMU04	Specchia (A)	39.9709 N; 18.2815 E
TMU05	Specchia (B)	39.9781 N; 18.1060 E
TMU06	Spinazzola	41.0072 N; 16.2658 E

A randomized block experimental design was used with four replicates having a plot area of 8 m^2. The macroplot sizes were 2 × 4 m, whereas microplot sizes were 1 × 1 m; each microplot was located in the centre of individual macroplots. Both the *T. mutabile* and ryegrass crops were fertilized uniformly with the solution of 10 kg N ha^{-1} ^{15}N NH$_4$SO$_4$ (10% ^{15}N atomic excess) (fertilizer dissolved in 1 L of distilled water). The macroplots also received 10 kg N ha^{-1} of unlabelled NH$_4$NO$_3$.

Forage biomass yield was determined by harvesting the whole plot at two main phenological stages including the vegetative stage and (stem elongation) and the regrowth at the flowering stage (inflorescence emergence-main shoot). The flowering time of the six accessions was also evaluated.

2.2. Nitrogen Fixation Analysis

Plant samples were oven-dried at about 70 °C for 48 h. The plant material was ground to pass through a 0.2 mm sieve. Total N and % ^{15}N atomic excess (a.e) of plant samples were analyzed at the Iso-Analytical Limited (Cheshire, UK) using elemental analysis isotope ratio mass spectrometry (EA-IRMS). The % ^{15}N excess was calculated by the difference of the atomic % ^{15}N in the plant material

(*T. mutabile* and ryegrass) and that of the natural abundance in the atmosphere (0.3663%). The N fixation in *T. mutabile* was calculated using the ^{15}N-isotope dilution method [10], using the equations (Equation (1)) mentioned below; and the amount of N_2 fixed was determined from the product % nitrogen derived from fixation (Ndfa) and N yield (as sum of the two cuts) for each replication, and the average was then calculated (Equation (2)):

$$\% \text{ Ndfa} = 100 \times [(1 - \% \, ^{15}\text{N a.e} \, (T. \, mutabile) / \% \, ^{15}\text{N a.e (ryegrass)}] \tag{1}$$

$$N_2 \text{ fixed (kg ha}^{-1}) = [(\% \text{ Ndfa} \times \text{total N in } T. \, mutabile \text{ (kg ha}^{-1})] / 100 \tag{2}$$

2.3. Forage Chemical Analysis

Samples of forage were ground in a hammer mill with a 1-mm screen and analyzed in triplicate for dry matter (DM), ash, crude protein, (CP; Kjeldahl N × 6.25), crude fiber and crude fat (ether extract with previous hydrolysis) according to the procedures described by the AOAC (Association of Official Analytical Chemists) [11]. Neutral detergent fiber (NDF) and acid detergent fiber (ADF) were determined according to Van Soest et al. [12], and were corrected for residual acid insoluble ash. Acid detergent lignin (ADL) was determined by the method of Van Soest and Robertson [13]. Representative samples of oven-dried forage weighing 1 g were placed in a muffle furnace at 550 °C for 4 h for total ash determination. The ash was wet with sulfuric and perchloric acids and diluted with distilled water [11]. Hemicellulose and cellulose also were estimated as NDF: (ash free)-ADF (ash free) and ADF (ash free)-ADL (ash free), respectively. All results were expressed on DM basis.

Digestibility of dry matter (DDM), dry matter intake (DMI), relative feed value (RFV), total digestible nutrients (TDN), net energy-lactation (NEl), net energy-maintenance (NEm) and net energy-gain (NEg) were calculated according to the following equations adapted from common formulas for forages [14]:

DDM % = 88.9 − (0.779 × ADF %);
DMI % = 120/NDF;
RFV = (DDM % × DMI %)/1.29;
TDN % = 96.35 − (ADF % × 1.15) (adapted equation from Lucerne formulas);
NEl: Mcal/1b = (TDN % × 0.01114) − 0.054;
NEm: Mcal/1b = (TDN % × 0.01318) − 0.132;
NEg: Mcal/1b = (TDN % × 0.01318) − 0.459.

2.4. Statistical Analysis

Data were analyzed by analysis of variance (ANOVA) using CoStat, version 1.03, Software (CoHort Software Incorporation, Monterey, CA, USA). The statistical analysis was applied to data following the one-way ANOVA design. Differences between means for significant effects were detected using Student-Newman-Keuls (SNK) test for biomass production and nitrogen fixation data. Unless otherwise stated, significance was declared at $p < 0.05$.

3. Results and Discussion

Unfortunately, previous studies evaluating the productive and nutritional characteristics including the nitrogen fixation have not yet been reported in *T. mutabile* forage. Thus, cross-referencing in the discussion of the findings in this study will be based on available results from other winter clover species under Mediterranean environmental conditions.

The six investigated accessions of *T. mutabile* led to a average total dry matter biomass production of 6.6 t/ha as sum of the two cuts at vegetative and flowering growth stages; the TMU06 and TMU02 accessions produced the higher biomass (~8.0 t/ha), whereas the two accessions TMU04 and TMU05 resulted in the lowest dry matter yield (5.5 and 5.1 t/ha, respectively) (Table 2). These results are comparable to those reported by Ovalle et al. [15] in *T. vesiculosum* in a high rainfall areas of the

Mediterranean climate zone of Chile and by Cazzato et al. [1] in different species of annual clovers in Southern Italy.

Table 2. Total aerial biomass yield, 1st cut biomass (% on total yield), flowering time, Ndfa % and fixed N of the different *T. mutabile* accessions.

Accession Code	Total Biomass Yield (t/ha)	Biomass Yield at 1st Cut (% on Total Yield)	Flowering Time	Ndfa * (%)	N$_2$ Fixed (kg/ha)
TMU01	6.2 bc	30.3 a	Early	64.5 cd	116.3 b
TMU02	8.0 a	23.6 b	Medium-Late	73.4 ab	191.2 a
TMU03	6.9 ab	15.0 c	Late	76.8 a	190.2 a
TMU04	5.5 bc	23.6 b	Medium-Early	58.2 d	96.3 b
TMU05	5.1 c	26.1 ab	Early	67.1 bc	111.4 b
TMU06	8.2 a	21.6 b	Medium	72.8 ab	186.0 a
Mean	6.6	23.4	-	68.8	148.6

* Ndfa: N derived from the atmosphere; Means within a column with no common letters (a–d) differ significantly ($p < 0.05$).

Considering the percentage of biomass yield at the first cut, it was found that the accessions produced 23.4% on average of the total biomass. In particular, the highest values were obtained by the early accessions (TMU01 and TMU05), conversely the lowest was detected in the late accession TMU03. Our findings confirm the data reported in the unique study conducted on *T. mutabile* in terms of regrowth capacity after forage cut at vegetative stage [6].

Based on flowering time, it was observed in our study a good variability for this parameter, with an average difference of ~20 days (from 150 to 170 days from emergence) between the earliest and latest accession (data not shown).

The Table 2 reports the values of N derived from air (Ndfa) and the N$_2$ fixed. The mean Ndfa was 68.8%, varying from 58.2% (TMU04) to 76.8% (TMU03). Regarding the N$_2$ fixed, the *T. mutabile* accessions showed an mean value of 148 kg/ha, with a significant variability among the six accessions (from 96 to 186 kg/ha). Our results on Ndfa and N$_2$ fixed are in agreement with those reported by Giambalvo et al. [16] in berseem clover (*T. alexandrinum*). Further, under similar environmental conditions, Cazzato et al. [1] found similar findings on Ndfa and N$_2$ fixed in many annual clovers (*T. incarnatum*, *T. resupinatum*, *T. squarrosum*, *T. alexandrinum* and *T. michelianum*).

The chemical composition of the different *T. mutabile* accessions at vegetative and flowering stages is reported in Table 3.

As expected, the forage cuts at the vegetative stage had a higher crude protein value (25.0%) compared to forage harvested at flowering stage (18.9%). A significant variation was observed among accessions varying from 22.8 to 29.4% of protein (first cut) and from 16.1 to 22.4% at second cut. Conversely, a higher fiber percentage was found at the second cut compared to the forage harvested at flowering stage (26.4 vs. 21.3%). Ross et al. [17] reported in berseem clover a crude protein declined between 31 to 18% DM according to the plant maturity.

The forage fat content did not vary significantly considering both the accessions and the two cuts (1.75% on dry matter). The ash content decreased on average from 12.7 (vegetative stage) to 10.3% (flowering stage), whereas the N-free extract increased harvesting the forage at flowering (39.9%) compared to vegetative (48.2%) stage.

The fiber fractions of the different *T. mutabile* accessions harvested at vegetative and flowering growth stages are reported in Table 4. Based on the findings, it was observed that all the determined parameters increased significantly from vegetative to flowering growth stage; in particular, the mean NDF content varied from 42.4 to 49.1%, the ADF from 23.3 to 29.7%, ADL from 6.7 to 7.1%, cellulose from 16.8 to 22.7%. However, taking into account the single accessions, a high variation in the fiber fractions evaluated was observed. The results of our accessions of *T. mutabile* for NDF and ADF values are similar to the findings reported by Evans and Mills [18] in arrowleaf clover. Moreover, considering the NDF and ADF values, as proposed by the Hay Marketing Task Force of the American

Forage and Grassland Council [19], the accessions of *T. mutabile* forage resulted of good quality when harvested at flowering stage and of "premium" quality when harvested at vegetative stage.

Table 3. Chemical composition of the different *T. mutabile* accessions at vegetative and flowering stages.

Accession Code	Vegetative Stage				
	Crude Protein (%)	Crude Fiber (%)	Crude Fat (%)	Ash (%)	N-Free Extract (%)
TMU01	29.4 (0.3)	16.1 (0.8)	1.6 (0.1)	12.5 (0.6)	40.3 (1.0)
TMU02	24.2 (0.1)	20.9 (0.4)	1.7 (0.1)	12.5 (0.1)	40.6 (0.9)
TMU03	24.1 (0.1)	22.7 (0.1)	1.9 (0.1)	12.5 (0.4)	38.7 (0.7)
TMU04	22.8 (0.1)	23.1 (0.2)	1.7 (0.1)	12.7 (0.2)	39.7 (0.2)
TMU05	23.1 (0.2)	22.2 (0.6)	1.9 (0.1)	12.4 (0.4)	38.5 (0.6)
TMU06	26.4 (0.1)	22.4 (0.4)	1.9 (0.1)	13.7 (0.6)	35.6 (1.2)
Mean	25.0	21.3	1.8	12.7	38.9
Significance level	$p < 0.01$	$p < 0.01$	ns	ns	$p < 0.01$
	Flowering Stage				
TMU01	22.4 (0.3)	23.7 (0.2)	1.7 (0.1)	10.4 (0.2)	41.8 (0.2)
TMU02	16.5 (0.1)	24.1 (0.4)	1.6 (0.1)	9.6 (0.3)	48.2 (0.8)
TMU03	17.7 (0.1)	29.2 (0.3)	1.6 (0.1)	10.3 (0.3)	41.3 (0.7)
TMU04	20.8 (0.1)	26.6 (0.4)	1.8 (0.1)	11.7 (0.3)	39.3 (0.1)
TMU05	19.4 (0.1)	28.3 (0.4)	1.9 (0.1)	10.8 (0.2)	39.6 (0.7)
TMU06	16.1 (0.1)	26.2 (0.2)	1.7 (0.1)	9.1 (0.3)	47.0 (0.2)
Mean	18.9	26.4	1.7	10.3	42.8
Significance level	$p < 0.01$	$p < 0.01$	ns	$p < 0.01$	$p < 0.01$

Values in brackets indicate the standard error; ns: not significant.

Table 4. Fiber fractions of the different *T. mutabile* accessions at vegetative and flowering stages.

Accession Code	Vegetative Stage				
	NDF (%)	ADF (%)	ADL (%)	Cellulose (%)	Hemicellulose (%)
TMU01	48.27 (0.49)	21.39 (0.52)	7.07 (0.08)	14.20 (0.48)	27.17 (0.72)
TMU02	42.25 (0.30)	23.34 (0.23)	6.80 (0.09)	17.14 (0.28)	18.52 (0.14)
TMU03	39.68 (0.28)	25.58 (0.11)	6.16 (0.04)	19.31 (0.04)	14.35 (0.42)
TMU04	41.52 (0.23)	22.60 (0.17)	6.58 (0.08)	16.46 (0.20)	18.93 (0.41)
TMU05	40.88 (0.33)	22.22 (0.25)	7.19 (0.07)	14.83 (0.11)	18.78 (0.20)
TMU06	41.60 (0.13)	24.78 (0.11)	6.14 (0.09)	18.79 (0.05)	16.29 (0.31)
Mean	42.42	23.31	6.70	16.84	19.00
Significance level	$p < 0.01$	$p < 0.01$	$p < 0.01$	$p < 0.01$	$p < 0.01$
	Flowering Stage				
TMU01	43.93 (0.28)	24.14 (0.18)	7.55 (0.19)	17.49 (1.28)	19.90 (0.21)
TMU02	50.03 (0.33)	28.50 (0.30)	7.18 (0.06)	21.18 (0.22)	22.01 (0.14)
TMU03	48.38 (0.26)	31.86 (0.46)	6.87 (0.02)	24.98 (0.43)	17.27 (0.04)
TMU04	52.46 (0.56)	29.11 (0.15)	6.99 (0.02)	22.15 (0.16)	23.77 (0.28)
TMU05	54.62 (0.42)	34.37 (0.59)	7.17 (0.19)	27.23 (0.36)	20.72 (0.64)
TMU06	45.46 (0.22)	30.04 (0.09)	6.99 (0.07)	23.01 (0.11)	15.58 (0.29)
Mean	49.11	29.73	7.12	22.71	19.94
Significance level	$p < 0.01$	$p < 0.01$	$p < 0.01$	$p < 0.01$	$p < 0.01$

NDF, neutral-detergent fiber; ADF, acid-detergent fiber; ADL, acid-detergent lignin; Values in brackets indicate the standard error.

Significant differences were observed among the six accessions of *T. mutabile* in terms of DDM, DMI and RFV for animal feeding (Table 5). As a result of differences in ADF contents, DDM contents of the accessions were also different in the two harvest period. Because of NDF contents of the accessions were different, also the DMI varied significantly. As a result of differences determined in DDM and DMI values amongst *T. mutabile* accessions, the RFV values were also different amongst accessions.

Table 5. Digestibility dry matter (DDM), dry matter intake (DMI) contents, relative feed value (RFV), Total digestible nutrients (TDN), net energy lactation (NEl), net energy. Maintenance (NEm) and net energy gain (NEg) contents of the different *T. mutabile* accessions.

Accession Code	Vegetative Stage						
	DDM	DMI	RFV	TDN	NEl	NEm	NEg
TMU01	72.23 (0.40)	2.49 (0.03)	139.23 (0.62)	71.75 (0.60)	0.75 (0.01)	0.82 (0.01)	0.49 (0.01)
TMU02	70.72 (0.18)	2.84 (0.02)	155.73 (0.70)	69.51 (0.27)	0.72 (0.0)	0.79 (0.01)	0.46 (0.01)
TMU03	68.98 (0.09)	3.03 (0.02)	161.71 (1.35)	66.94 (0.13)	0.69 (0.0)	0.75 (0.0)	0.42 (0.0)
TMU04	71.30 (0.14)	2.89 (0.02)	159.72 (0.61)	70.36 (0.20)	0.73 (0.0)	0.80 (0.01)	0.47 (0.0)
TMU05	71.59 (0.20)	2.94 (0.02)	162.91 (1.77)	70.79 (0.29)	0.74 (0.01)	0.80 (0.0)	0.48 (0.01)
TMU06	69.60 (0.08)	2.89 (0.01)	155.64 (0.31)	67.86 (0.12)	0.70 (0.0)	0.76 (0.0)	0.44 (0.01)
Mean	70.71	2.84	155.82	69.53	0.72	0.78	0.46
Significance level	$p < 0.01$	$p < 0.01$	$p < 0.01$	$p < 0.01$	$p < 0.01$	$p < 0.01$	$p < 0.01$
	Flowering Stage						
TMU01	70.09 (0.14)	2.73 (0.02)	148.43 (1.24)	68.59 (0.21)	0.71 (0.01)	0.77 (0.01)	0.44 (0.01)
TMU02	66.70 (0.23)	2.39 (0.01)	124.04 (0.38)	63.58 (0.34)	0.66 (0.0)	0.71 (0.01)	0.38 (0.01)
TMU03	64.08 (0.35)	2.48 (0.01)	123.23 (3.03)	59.71 (0.52)	0.61 (0.0)	0.66 (0.01)	0.33 (0.01)
TMU04	66.22 (0.11)	2.28 (0.02)	117.45 (1.03)	62.87 (0.17)	0.64 (0.0)	0.69 (0.01)	0.37 (0.01)
TMU05	62.13 (0.46)	2.19 (0.01)	105.83 (1.58)	56.83 (0.67)	0.58 (0.01)	0.62 (0.01)	0.29 (0.01)
TMU06	65.50 (0.06)	2.64 (0.01)	134.04 (0.78)	61.81 (0.1)	0.63 (0.01)	0.68 (0.01)	0.36 (0.01)
Mean	65.82	2.46	125.52	62.23	0.64	0.69	0.36
Significance level	$p < 0.01$	$p < 0.01$	$p < 0.01$	$p < 0.01$	$p < 0.01$	$p < 0.01$	$p < 0.01$

Values in brackets indicate the standard error.

Differences in terms of TDN, NEl, NEm and NEg contents were obtained amongst accessions. The relative feed value (RFV) is a widely accepted forage quality index in the marketing of hays in the United States of America [20,21]. Based on our results on RFV, the forage obtained from the *T. mutabile* accessions can be classified as good-premium quality, resulting thus a valuable feeding resource for livestock species.

4. Conclusions

From our findings, significant differences were found among the accessions of *T. mutabile* in terms of biomass production, capacity of nitrogen fixation and forage quality characteristics. Besides the high forage yield, the collected accessions showed favourable values of production and quality, representing also valuable germplasm for further selection programs as well as the application for possible plant cultivar registration. Moreover, it is interesting to underline that *T. mutabile* may represent a valuable alternative to commonly cultivated annual clover species especially in Mediterranean areas, because of its longer vegetative cycle. However, further investigations are needed to assess the self-reseeding capacity of *T. mutabile* when utilized as pasture species.

Author Contributions: Conceptualization, E.C. and M.F.; Methodology, E.C. and M.F.; Formal Analysis, C.L.; Writing—Original Draft Preparation, E.C. and M.F; Writing—Review and Editing, V.L.

Funding: This research was funded by the project SaVeGraIN Puglia – Progetto Integrato per la Biodiversità, PSR Regione Puglia, FEASR 2014–2020, Misura 10.2.1 "Progetti per la conservazione e valorizzazione delle risorse genetiche in agricoltura e selvicoltura" Regulation European Community 1305/2013 trascinamento Misura 214 PSR Puglia 2007/2013.

Acknowledgments: The authors want to thank the field technical support of Mario Alberto Mastro involved in the present study.

References

1. Cazzato, E.; Annese, V.; Corleto, A. Azoto fissazione in leguminose foraggere annuali in ambiente mediterraneo. *Rivista di Agronomia* **2003**, *37*, 57–61.

2. Puckridge, D.W.; French, R.J. The annual legume pasture in cereal-Ley farming systems of southern Australia: A review. *Agric. Ecosyst. Environ.* **1983**, *9*, 229–267. [CrossRef]

3. Scoppola, A.; Lattanzi, E.; Bernardo, L. Distribution and taxonomy of the Italian clovers belonging to *Trifolium* sect. *Vesicastrum* subsect. *Mystillus* (Fabaceae). *Ital. Bot.* **2016**, *2*, 7–27. [CrossRef]

4. Greuter, W.; Burdet, H.M.; Long, G. (Eds.) *Conservatoire et Jardin Botaniques*; Med-Checklist 4; Conservatoire et Jardin Botanique Ville de Genève: Genève, Switzerland, 1989; p. 458.

5. Conti, F.; Abbate, G.; Alessandrini, A.; Blasi, C. (Eds.) *An Annotated Checklist of the Italian Vascular Flora*; Palombi Editori: Roma, Italy, 2005; p. 420.

6. Hoveland, C.S.; Alison, M.W. *RyeeRyegrasseLegume Trials in Alabama 1978–1981*; Bull. 543; Alabama Agricultural Experiment Station, Auburn University: Auburn, AL, USA, 1982; p. 12.

7. Wilman, D.; Altimimi, M.A.K. The in-vitro digestibility and chemical composition of plant parts in white clover, red clover and lucerne during primary growth. *J. Sci. Food Agric.* **1984**, *35*, 133–138. [CrossRef]

8. McGraw, R.L.; Marten, G.C. Analysis of Primary Spring Growth of Four Pasture Legume Species[1]. *Agron. J.* **1986**, *78*, 704–710. [CrossRef]

9. Lloveras, J.; Iglesias, I. Morphological development and forage quality changes in crimson clover (*Trifolium incarnatum* L.). *Grass Forage Sci.* **2001**, *56*, 395–404. [CrossRef]

10. Hardarson, G.; Danso, S.K.A.; Zapata, F.; Reichardt, K. Measurements of nitrogen fixation in faba bean at different N fertilizer rates using the ^{15}N isotope dilution and 'A-value' methods. *Plant Soil* **1991**, *131*, 161–168. [CrossRef]

11. AOAC. *Official Methods of Analysis*, 17th ed.; Association of Official Analytical Chemists: Arlington, VA, USA, 2000.

12. Van Soest, P.J.; Robertson, J.B.; Lewis, B.A. Methods for dietary fiber, neutral detergent fiber, and non-starch polysaccharides in relation to animal nutrition. *J. Dairy Sci.* **1991**, *74*, 3583–3597. [CrossRef]

13. Van Soest, P.J.; Robertson, J.B. Analysis of forages and fibrous feeds. In *Laboratory Manual for Animal Science 613*; Cornell University: Ithaca, NY, USA, 1985; p. 18.

14. Schroeder, J.W. Interpreting Forage Analysis. Extension Dairy Specialist (NDSU), AS-1080, North Dakota State University. May 1994. Available online: https://library.ndsu.edu/ir/bitstream/handle/10365/9133/AS-1080-1994.pdf?sequence=2 (accessed on 10 January 2018).

15. Ovalle, C.; del Pozo, A.; Fernández, F.; Chavarría, J.; Arredondo, S. Arrowleaf clover (*Trifolium vesiculosum* Savi): A new species of annual legumes for high rainfall areas of the Mediterranean climate zone of Chile. *Chil. J. Agric. Res.* **2010**, *70*, 170–177. [CrossRef]

16. Giambalvo, D.; Ruisi, P.; Di Miceli, G.; Frenda, A.S.; Amato, G. Forage production, N uptake, N_2 fixation, and N recovery of berseem clover grown in pure stand and in mixture with annual ryegrass under different managements. *Plant Soil* **2011**, *342*, 379–391. [CrossRef]

17. Ross, S.M.; King, J.R.; O'Donovan, J.T.; Spaner, D. The productivity of oats and berseem clover intercrops. I. Primary growth characteristics and forage quality at four densities of oats. *Grass Forage Sci.* **2005**, *60*, 74–86. [CrossRef]

18. Evans, P.M.; Mills, A. Arrowleaf clover: Potential for dryland farming systems in New Zealand. In *Proceedings of the New Zealand Grassland Association*; New Zealand Grassland Association: Wellington, New Zealand, 2008; Volume 70, pp. 239–243.

19. Rohweder, D.A.; Barnes, R.F.; Jorgensen, N. Proposed hay grading standards based on laboratory analyses for evaluating quality. *J. Anim. Sci.* **1987**, *47*, 747–759. [CrossRef]

20. Tavlas, A.; Yolcu, H.; Tan, M. Yields and qualities of some red clover (*Trifolium pratense* L.) genotypes in crop improvement systems as livestock feed. *Afr. J. Agric. Res.* **2009**, *4*, 633–641.

21. Tucak, M.; Popovic, S.; Cupic, T.; Spanic, V.; Meglic, V. Variation in yield, forage quality and morphological traits of red clover (*Trifolium pratense* L.) breeding populations and cultivars. *Zemdirbyste Agric.* **2013**, *100*, 63–70. [CrossRef]

agriculture

MDPI

Review

Effect of Forage *Moringa oleifera* L. (moringa) on Animal Health and Nutrition and Its Beneficial Applications in Soil, Plants and Water Purification

Mohamed E. Abd El-Hack [1], Mahmoud Alagawany [1,*], Ahmed S. Elrys [2], El-Sayed M. Desoky [3], Hala M. N. Tolba [4], Ahmed S. M. Elnahal [5,6], Shaaban S. Elnesr [7] and Ayman A. Swelum [8,9]

[1] Department of Poultry, Faculty of Agriculture, Zagazig University, Zagazig 44511, Egypt;
     dr.mohamed.e.abdalhaq@gmail.com
[2] Department of Soil Science, Faculty of Agriculture, Zagazig University, Zagazig 44511, Egypt;
     elrys_sms2008@yahoo.com
[3] Department of Agriculture Botany, Faculty of Agriculture, Zagazig University, Zagazig 44511, Egypt;
     desoky_s@yahoo.com
[4] Department of Avian and Rabbit Medicine, Faculty of Veterinary Medicine, Zagazig University,
     Zagazig 44511, Egypt; Moonfacem2000@yahoo.com
[5] College of Plant Protection, Northwest A&F University, Yangling 712100, China;
     abo_mariam2017@yahoo.com
[6] Department of Plant Pathology, Faculty of Agriculture, Zagazig University, Zagazig 44511, Egypt
[7] Department of Poultry Production, Faculty of Agriculture, Fayoum University, Fayoum 63514, Egypt;
     ssn00@fayoum.edu.eg
[8] Department of Animal Production, College of Food and Agriculture Sciences, King Saud University,
     P.O. Box 2460, Riyadh 11451, Saudi Arabia; aswelum@ksu.edu.sa
[9] Department of Theriogenology, Faculty of Veterinary Medicine, Zagazig University, Zagazig 44511, Egypt
*   Corresponding: mmalagwany@zu.edu.eg or dr.mahmoud.alagawany@gmail.com; Tel.: +20-114-300-3947

Received: 7 August 2018; Accepted: 11 September 2018; Published: 18 September 2018

Abstract: *Moringa oleifera* L. (moringa) is known as one of the most useful multipurpose plants. It can be effectively utilized as a natural biopesticide and inhibitor of several plant pathogens. Thus, it can be included in integrated pest management strategies. Moringa and its products have different uses in many agricultural systems. The use of moringa as a crop enhancer is an eco-friendly way of improving crop yields at the lowest possible cost. This inexpensive increase in productivity can contribute to meeting some of the food needs in some parts of the world as the global population increases and poverty rates rise. One of the most important characteristics of moringa is that it has high biological and nutritional values and can be used as animal feed, green fertilizer, medicine, biopesticide and in seed production. Moringa has been characterized as a potentially useful animal feed owing to its high content of protein, carotenoids, several minerals and vitamins (such as iron and ascorbic acid) and certain phytochemicals (kaempferitrin, isoquercitrin, rhamnetin, kaempferol and quercetin). This review aims to provide more knowledge about the nature, nutritional value, phytochemicals and uses of *Moringa oleifera* as a promising material in the fields of soil and plant management, water treatment, as well as animal and poultry production.

Keywords: *Moringa oleifera*; forage; beneficial use; nutritional composition; poultry; animal; plant

1. Introduction

In recent years, different phytogenic feed additives, i.e., aromatic plants or their respective essential oils have been investigated for poultry [1–3]. Different species have been tested at various inclusion levels to find a cheap, safe and natural feed additive with high economic output [4]. The use of phytogenic compounds in feed might improve feed quality. Firstly, they possess the ability to retard the growth of mycotoxigenic fungi [5]. Secondly, their antibacterial and antioxidative properties (e.g., thymol, carvacrol and rosmarinic acid) contribute to improving the overall quality of feed [6]. Antioxidant properties could be of great importance when the feed contains a high proportion of polyunsaturated fatty acids, versus saturated fatty acids. Phytogenic extracts/essential oils possess antimycotic properties, which could be helpful in preventing mycotoxin production in stored wheat grains. Growth of toxigenic fungi, e.g., *Aspergillus flavus*, *Aspergillus parasiticus*, *Aspergillus ochraceus* and *Fusarium moniliforme*, could be inhibited by anise, thyme and cinnamon [7,8]. Overuse and misuse of pesticides and fungicides to manage pests and plant pathogens have resulted in harmful effects including death, birth defects among humans and animals and several diseases (such as cancer, allergies, etc.) [9]. Due to the negative impacts associated with pesticide usage, much attention has been focused on alternative ecofriendly methods of pathogen control. There is an urgent need to examine non-synthetic chemical approaches to disease management in agricultural applications [10,11].

Moringa oleifera L. (moringa), belongs to family Moringaceae, grows in tropical and subtropical environments. Every part of the tree has beneficial properties, making it multipurpose tree where is used as herbal medicine, spices, food, fertilizer natural coagulants, forage and nectar for bees. Additionally, it is used as good sources of vitamins (A, B and C), nicotinic acid, riboflavin, pyridoxine, folic acid, beta-carotene, ascorbic acid, alpha-tocopherol calcium and iron as well as a main source of the essential amino acids [12]. Reports have also described the plant to be anti-inflammatory, antioxidant, antimicrobial and antitumor activity. Moringa has powerful antibiotic and fungicidal effects [13]. As well, moringa has the potential to improve nutrition and support immune functions of poultry and animal, where responses to moringa include reduced *Escherichia coli* and increased *Lactobacillus* counts in the intestine demonstrating an enhanced immune response [14].

Moringa oleifera has other agricultural uses beyond supplementation for animals. There are many reports that moringa leaf extract (MLE) can play a major role in accelerating the growth of tomatoes, peanuts, maize and wheat in the early vegetative stages. Also, *Moringa oleifera* (ethanol or aqueous) extracts are regarded as biopesticidal products that are ecofriendly, have low cash input, are readily available, have a minimal environmental impact and are helpful in plant disease management. *M. oleifera* is reported to have antimicrobial properties against plant pathogens that cause serious plant diseases, such as soil-borne disease [15]. Moringa can also be effectively used as a natural biopesticide and, thus, it can be included in integrated pest management (IPM) strategies [16]. A significant improvement in pest and disease resistance has also been observed with MLE use, with overall yield increases of 20% to 35% [17]. The use of MLE as a crop enhancer is an environmentally friendly way of increasing crop yields at the lowest possible cost. This inexpensive increase in productivity can contribute to meeting some of the food needs in some parts of the worldwide, associated with the global population increase, as poverty rates rise [18]. One of the most important characteristics of *M. oleifera* is that it has a high nutritional value and can be used in animal feed, green fertilization, medicines, biopesticides and seed production [17]. This article aims to provide more knowledge about the nature, description, nutritional values and uses of *M. oleifera* as a promising material in soil and plant management, water treatment and the animal and poultry industry.

2. Description of *Moringa oleifera*

Moringa oleifera is commonly termed the "drumstick tree". Other common names include horseradish tree, ben oil tree, or benzoil tree. Some parts of moringa tree (leaves, pods, seeds, flowers, fruits and roots) are eaten as food and some are taken as a remedy. *Moringa oleifera* is a fast-growing, deciduous tree. Its maximum height is 10–12 m, while its trunk can reach a diameter of 45 cm. The flowers are approximately 1.0–1.5 cm long and 2.0 cm wide. Flowering starts within the first six months after planting. The fruit is a droopy, three-sided brown capsule, 20–45 cm in size and contains dark brown, spherical seeds of about 1 cm diameter. The seeds have three thin, whitish wings, which are responsible for the smooth distribution of the seed by water and wind [19]. This tree also needs a yearly rainfall of between 250 mm and 3000 mm and can survive in temperatures of 25 °C to 40 °C, which makes it suitable for tropical climates. Moringa tree grows mainly in semi-arid tropical and subtropical areas. More generally, moringa grows in the wild or is cultivated in Central America and Caribbean, northern countries of South America, Africa, Southeast Asia and various countries of Oceania. Among the twelve species in the genus *Moringa*, the most commonly cultivated and widespread is *Moringa oleifera* [20], which is native to the sub-Himalayan tract of India and Pakistan [21].

2.1. Nutritional Composition of the Moringa oleifera

All parts of *Moringa oleifera* are consumed by humans in different ways. Moringa leaves are a source of highly digestible nutrients and can be eaten fresh, cooked or stored as dried powder and have been advocated as suitable for nutritional and therapeutic use in many developing regions of the world [22]. Moringa has a great possible in improving nutrition and support immune functions of poultry and animal. Seeds are eaten green or dry [23]. Moringa seeds contain a high percentage of sweet oil (30–40% of the seed weight) and contain around 76% polyunsaturated fatty acids which can control cholesterol. The leaves and seeds of *Moringa oleifera* are a source of protein, iron, calcium, ascorbic acid vitamin A and antioxidant compounds such as carotenoids, flavonoids, vitamin E and phenolics [24]. The presence of vitamins and minerals benefit in improving the immune system and cure a myriad of diseases [25]. Various amino acids, such as Arg, His, Lys, Trp, Phe, Thr, Leu, Met, Ile, Val are present in *Moringa oleifera* leaves [25].

Moringa oleifera leaves can be used as a feed supplement, to improve feed efficiency and livestock performance, or as a replacement for conventional crops to obtain more economically sustainable, environmentally friendly and safer production [26,27]. Rubanza et al. [28] reported better feed digestibility in animals fed moringa leaves, probably due to their nutritional profile especially the neutral detergent fiber (NDF), acid detergent fiber (ADF), crude protein (CP), gross energy (GE), ether extract (EE) and amino acids. Some parts of the moringa tree contain toxins and other anti-nutritional factors, which limit their utility as a source of food for humans or animals. The bark of the tree contains alkaloids, tannins, saponins and some inhibitors [29,30]. Grubben and Denton [31] reported two types of alkaloids in moringa root bark, i.e., moringinine and moringine. Bose [32] stated that bark of moringa tree has toxicity profiles due to its content of two alkaloids and the toxic hypotensive moringinine. The bark of the tree may cause violent uterine contractions that can be fatal [33]. Faizi et al. [34] reported that using of Niazinin A, niazimicin resulted from the ethanolic extract of *Moringa oleifera* leaves produced negative inotropic and chronotropic effects in isolated guinea pig atria and hypotensive and bradycardiac effects in anesthetized rats. Additionally, changes in clotting factor, changes in serum composition (e.g., total protein, bilirubin, cholesterol), along with enzyme inhibition, induction, or change in blood or tissue levels of other transferases have been noted after the mice treatment by root bark extract *Moringa oleifera* Lam. are 500 mg/kg and 184 mg/kg. Even though the toxic root bark is removed, the flesh has been found to contain the alkaloid spirochin, which can cause nerve paralysis [35]. Mazumder et al. [36] found that methanolic extract of *Moringa oleifera* root was found to contain 0.2% alkaloids and high doses (>46 mg/kg body weight) of crude extract affect liver and kidney function and hematologic parameters. Phytate content in *M. olifera* leaves is about 3.1%, which might decrease availability of minerals in monogastrics [30]. A point to note is that the nutrient composition varies depending on the climate, location and the environmental factors significantly influence nutrient content of the tree [25]. Finally, we can conclude that the plant can be dangerous if consumed too frequently or in large amounts. Therefore, it has been suggested that more attention should be paid to how the plant has been used in diets.

2.2. Phytochemicals of the Moringa oleifera

Phytochemical studies of moringa leaves have indicted unique compounds, including rhamnose (i.e., simple sugar), isothiocyanate and glucosinolates [13,37] that are known for strong hypotensive (blood pressure lowering) and spasmolytic (muscle relaxant) effects [38]. Other important compounds such as benzyl glucosinolates, 4-(4-*O*-acetyl-α-L-rhamnopyranosyl oxy) benzyl thiocyanate and 4-(α-L-rhamnopyranosyl oxy) benzyl isothiocyanate are also present. These compounds are known to possess anticancer, hypotensive and antibacterial activity [13]. Some flavonoid pigments, such as kaempferitrin, isoquercitrin, rhamnetin, kaempferol and quercetin are found in moringa flowers [13,37]. Cytokine-type hormones were observed in MLE in 80% ethanol [29,39]. Al-Asmari et al. [40] reported that such ethanol extracts had cancer preventative effects, when they assayed the activity against human promyelocytic leukemia cells (HL-60).

According to Yameogo et al. [41], *Moringa oleifera* is the best source of a wide spectrum of dietary antioxidants, including flavonoids such as kaempferol and quercetin. Siddhuraju and Becker [42] reported the concentrations of natural antioxidants in *Moringa oleifera* from three different agroclimatic origins: on a dry weight basis, phenolics = 74–210 µmol g^{-1}, ascorbate (vitamin C) = 70–100 µmol g^{-1}, β-carotene = 1.1–2.8 µmol g^{-1} and α-tocopherol = 0.7–1.1 µmol g^{-1}. Notably, they had higher antioxidant contents than fruits and vegetables that are known for their high antioxidant contents, e.g., strawberries (high in phenolics ~190 µmol gallic acid (GA) g^{-1}), carrots (high in β-carotene ~1.8 µmol g^{-1}), soybean (high in α-tocopherol ~1.8 µmol g^{-1}) and hot pepper (high in ascorbate ~110 µmol g^{-1}) [43]. As well, Pakade et al. [44] stated that moringa excels some vegetables in their strength as an antioxidant, because that its content of total phenolics was almost twice that of the vegetables (broccoli, spinach, peas and cauliflower) and total flavonoids were three times that of the same vegetables. Also, the reducing power of moringa was higher and free radicals remaining were lower compared with these vegetables. Moringa leaves also contain other important flavonoids such as kaempferol and quercetin, which exhibit higher antioxidant activity than ascorbic acid [40,45]; they also contain fairly high amounts of ascorbic acid [42]. These antioxidants provide protection to animals against degenerative diseases and infections, which might be associated with the direct trapping of free radicals to avoid DNA damage from excessive oxidation [46,47]. Moringa seeds contain ferulic acid, gallic acid, epicatechin, catechin, vanillin, protocatechuic acid, caffeic acid, cinnamic acid, phytosterol, quercetin, chlorogenic acid and quercetin rhamnoglucoside. The immature pods (fruits) and flowers of *Moringa oleifera* have been characterized by the content of carotenoids. Also, isothiocyanate from moringa seeds acts as anticancer agents and mitigates oxidative stress [25,48]. Phytosterols such as kampesterol sitosterol and stigmasterol are precursors for hormones, where induce the production of estrogen, subsequently, stimulates the proliferation of the mammary gland ducts to create milk. Additionally, the presence of flavanoids give the anti-inflammatory, antioxidant and antidiabetic properties and as anti-proliferative and anticancer agent [25].

The phytochemical compounds of moringa possess various biological actions, including antidiabetic, hypocholesterolemic, hypertensive agent and regulate thyroid hormone, central nervous system, digestive system, as well as nutrition and metabolism. Reports have also described the plant to be highly potent anti-inflammatory agent anti-inflammatory, antimicrobial and antitumor activity. Finally, we can state that moringa is rich in phytochemical compounds that confer on the plant significant medicinal properties that could be valuable for treating certain ailments.

3. Uses of Moringa in Soil and Plants

The moringa tree is one of the most nutrient-rich plants in the world. It has many uses for plant and soil such as a green manure and natural growth stimulants. Exogenous application of MLE, whether it is an aqueous or ethanol extract, improves productivity in many crops, because MLE possesses great antioxidant activity and is rich in plant secondary metabolites such as ascorbic acid and total phenols, making it a potential natural growth stimulant [49]. Several studies have focused on the role of MLE in improving plant growth and increasing the production of numerous crops [49,50]. Moreover, MLE, such as other bio-stimulants, is used to enhance plant resistance to abiotic stresses [51].

Moringa is a plant that grows under drought conditions and in all soil types [52]. Using moringa shoot as a green manure can significantly enrich agricultural soil. In this method, the soil is first plowed and moringa seed is then planted 2 cm deep at a spacing of 10×10 cm (a density of 1 million seeds per hectare). The seedlings are plowed into the soil, to a depth of 15 cm and then the soil is prepared for the desired crop [53–55]. Fresh moringa leaf aqueous extracts contain many antioxidants and is rich in secondary metabolites and osmoprotectants [51,56]. In addition, MLE is a source of vitamins, zeatin, indole-3-acetic acid (IAA), cytokinin, gibberellins (GAs) and several mineral elements (i.e., P, Ca, K, Mg, Fe, Cu, Zn and Mn) [56,57]. The various ingredients contained in the MLE suggest that this aqueous extract can be used effectively as a plant biostimulant and, thus, is considered to be one of the most natural growth stimulants available [57]. This tree also contains numerous curative properties and chemical materials for other uses and is sometimes called "the prodigy tree" [58]. These advantages are linked to the geographical distribution of these trees; they are particularly valuable as they are found in areas with high population density and high poverty rates [58]. Among several uses of moringa is use of the leaf water extract as a hormone promoter for numerous crops [59,60]. Therefore, MLE as growth promoter can be a natural and practical alternative supplement to synthetic sources applied to improve productivity in crop plants. In addition, it benefits the plant and the soil together for their effective advantages resulting from its content of many bioactive components.

3.1. Effect of Moringa oleifera on Plant Growth Characteristics and Yield

Moringa leaf ethanol extract has plant-growth-promoting capabilities as it is rich in K, Ca, carotenoids, phenols and zeatin [29]. Three sprays of MLE significantly influenced the number of branches/plant, plant height, number of pods/plant, number of seeds/pod and biological yield of canola, compared to untreated control plants [59]. Many studies have demonstrated that the addition of MLE enhanced crop yield, as determined by strong seedling growth and yield of bean (*Phaseolus vulgaris* L.) [56,57]. The different concentrations of moringa extract was capable to enhance the photosynthetic apparatus in treated plant, which leads to increase in plant productivity and fruit dry matter [61]. Enhanced seedling growth parameters (number and area of leaves/plant, shoot length, dry weight of plant, reducing sugars, amylase activity and plant growth) following the addition of MLE may be due to the mobilization of germination-related metabolites/inorganic solutes (i.e., ascorbic acid, zeatin, Ca and K) in the growing plumule [29,62,63]. Moreover, the high content of GAs, IAA and zeatin in MLE promotes linola (*Linum usitatissimum* L.) plant growth and production under saline conditions in comparison to untreated and hydropriming controls [50]. The rapid increase in growth observed in these studies is likely due to the enriched content of crude proteins as well as auxins and cytokinins, which are growth-promoting hormones [64]. Proteins are fundamental for the formation of the protoplasm; however, growth hormones promote fast cell division, cell multiplicity and cell enlargement [61]. A significant reduction in growth (in terms of shoot and root length and plant dry mass), after treatment with 100 mM NaCl (compared to a control, distilled water), which resulted in a significant loss in the yield of common bean (*Phaseolus vulgaris* L.) plant [51]. However, pre-soaking the beans in MLE for 8 h significantly increased the growth parameters. The combined treatment of MLE + NaCl (100 mM) alleviated the adverse effects of NaCl-salinity and maintained the growth traits and bean yield at the same levels as the control plants. Cytokinin is one of the most important growth regulators discovered in MLE. It has an important role in increasing the chlorophyll content of plants and improving cell division [65].

The growth parameters (i.e., number of leaves/plant, shoot length and dry weight) of common bean plants grown under saline conditions were positively affected by the application of MLE in two seasons. Addition of MLE, as a seed soak (SS) or foliar spray (FS), led to significant increases in all growth characteristics compared to control plants (Figure 1(1)) [66,67]. A plant growth spray made from moringa leaves enhanced crop production by 20–35% [29]. In addition, numerous experiments have shown that spraying plants with MLE accelerated plant growth [68,69]. Howladar [56] reported that NaCl and/or CaCl2 application significantly decreased the growth parameters and yield of bean plants. It was found that the combined influence of these two stress factors was more harmful than their individual affects. Compared to the control plants, the shoot length, root length, leaf area and plant dry weight decreased by 28.1%, 49.6%, 40.2% and 63.2% respectively. In addition, the number of pods/pot, pod protein and pod yield/pot were reduced by 62.7%, 63.3% and 51.2% respectively. Plant treatment with MLE, in the absence of NaCl and/or CaCl₂, stimulated bean growth and yield and were significantly higher than in the control group. Foliar application of MLE also enhanced the growth of bean plants after treatment with NaCl and/or CaCl₂ and the values were significantly higher than those of the plants grown under stress alone [56].

Moringa leaf ethanol extract is a natural plant growth enhancer, has low cost and enhances the tolerance of plants for difficult environmental conditions, such as drought. However, MLE has also gained attention owing to its high content of proteins, antioxidants (ascorbic acid, flavonoids, phenolics and carotenoids), mineral ions (P, Ca, Fe, K, Cr, Cu, Mg, Mn and Zn), amino acids, vitamin A, vitamin C, B-complex and plant hormones, especially cytokinins (zeatin) [70]. Drought stress has a destructive effect on the cytokinin content of plants; therefore, the high levels of zeatin in MLE make it particularly effective as a natural compound promoting plant tolerance under stress conditions [61]. Hanafy [71] observed that MLE is rich in nutrients, antioxidants (such as ascorbic acid, α- tocopherol, phenols and flavonoids), as well as phytohormones (such as IAA and GAs). Plant growth traits (fresh and dry weights of shoots and roots, shoot and root length) of *Glycine max* plants were affected by drought and the effect increased as the water holding capacity of the plants decreased (60% and 40%). It was reduced significantly with increasing drought stress as compared with the control group. Spraying plants with MLE resulted in a noticeable improvement in all growth parameters, compared to untreated plants, indicating higher growth efficiency and development with MLE supplementation. Plants treated with MLE and subjected to drought stress showed highly significant increases in growth traits compared with either drought-stressed plants or well-watered plants.

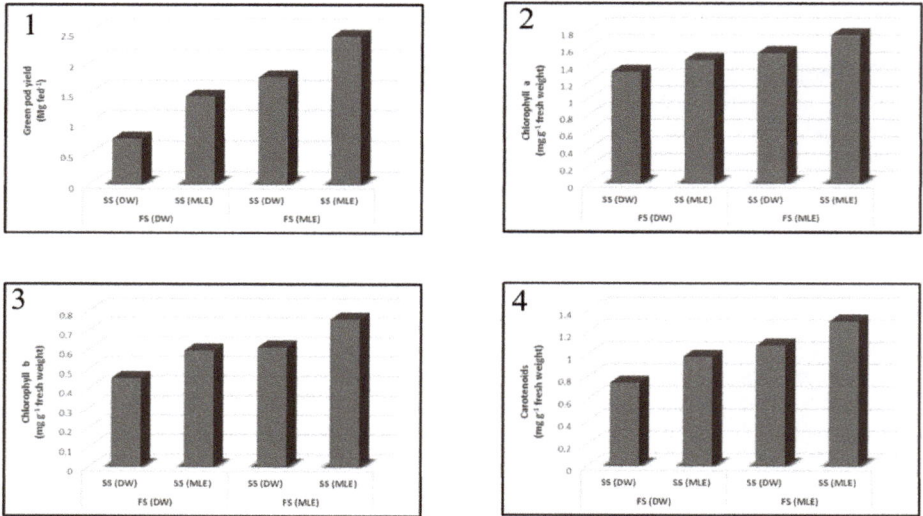

Figure 1. *Cont.*

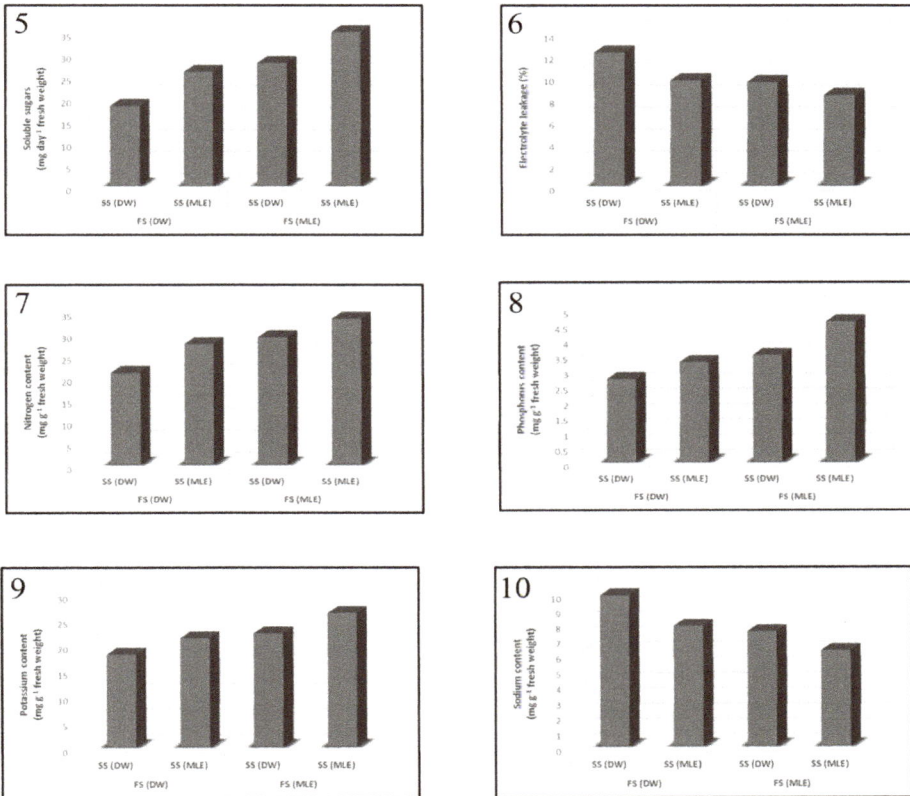

Figure 1. Effect of foliar spray (FS) and/or seed soaking (SS) with MLE or distilled water (DW) on (**1**) green pod yield, (**2**) chlorophyll a, (**3**) chlorophyll b, (**4**) carotenoids, (**5**) soluble sugars, (**6**) electrolyte leakage, (**7**) nitrogen content, (**8**) phosphorus content, (**9**) potassium content and (**10**) sodium content of Pea (*Pisum sativum* L.) plants growing under salinity stress condition [66].

3.2. Effect of Moringa oleifera on Physicochemical Attributes

All physicochemical attributes (i.e., total chlorophyll, total carotenoids, relative water content (RWC, %), electrolyte leakage (EL, %), membrane stability index (MSI, %), free proline, total soluble sugars, ascorbic acid, N, K, P, Ca, K/Na ratios, Ca/Na ratios, K + Ca/Na ratios and antioxidant enzymes) (Figure 1(2–10)) of pea (*Pisum sativum* L.) were affected by the application of MLE under salinity stress conditions [61,66]. Higher chlorophyll content and enhanced growth traits were positively reflected in the dry seed yield and green pods of bean plants grown under saline conditions. These increases may be due to assimilation of ascorbic acid and have been linked to cytokinin levels found in MLE [51,72]. Salt stress significantly increased EL; however, the addition of MLE significantly reduced it. This decrease in EL was greater when MLE was applied as both a SS and FS [61,66]. When plants are subject to salinity stress, EL (Figure 1(6)) can cause damage to cell membranes. Maintaining the integrity of cell membranes under saline conditions is an important goal of salinity tolerance techniques [73]. The addition of MLE significantly enhanced the RWC of plants, compared to the control plants and the highest elevation in RWC was observed when MLE was applied as a SS combined with a FS. Relative water content is a useful measure of the physiological condition of plants [74]. MLE is reported to enhance RWC and the tolerance of plants to salt stress, probably due to increases in the concentrations of osmoprotectants [66,67].

The use of an MLE foliar spray on bean plants grown under saline conditions led to a significant increase in soluble sugar content (Figure 1(5)) [51,66]. The MLE participates in osmotic adaptation and ability by directly, or indirectly, modifying the code of genes involved in metabolic procedures, storage functions and defense [75]. It has been reported that the oxidative injury created during salinity stress is due to a disparity in the production of reactive oxygen species (ROS) and antioxidant active alterations [76]. To avoid oxidative stress injury, plants have evolved antioxidant systems, including non-enzymatic ones, such as ascorbic acid (AsA), that act to directly lower ROS during various forms of stress [77]. In this respect, the maximum concentrations of AsA in bean plants were found in MLE-treated plants [78]. Since AsA can immediately remove O_2 and H_2O_2 through a non-enzymatic path, the addition of foliar MLE can be used to inhibit O_2 accumulation [79]. The presence of zeatins, such as cytokinin, in MLE prevents early aging of the leaves and preserves the largest possible leaf area, thereby increasing the rate of photosynthesis [57]. Cytokinin levels are usually reduced in the later stages of plant growth. Therefore, the external addition of cytokinins, through the application of MLE, can delay this process and increase the chlorophyll and soluble sugar content in salt-stressed bean plants [57,80]. This study demonstrated that the inhibitory effects of abiotic stress on the growth and production of plants could be mitigated by the exogenous addition of MLE.

3.3. Applications of Moringa oleifera in Water Treatment and Purification

Softening is the abstraction of ions that cause hardness in water. Hardness is mostly caused by Ca and Mg ions, or at times, by Fe, Mn, strontium and Al ions. Hardness can lead to excessive soap consumption. In general, water hardness should not be above 300 mg L^{-1} to 500 mg L^{-1}; hardness greater than 150 mg L^{-1} may require softening [81]. Chemicals used for water treatment are expensive and are not available locally in most developing countries. Thus, biological anticoagulants that can be used in water desalination, such as *Moringa oleifera* seeds, need to be investigated [82]. Moringa is one of the most important natural substances that can be used in the purification of drinking water [22,81] at low cost and low risk to human health and the environment [83]. Dried, ground moringa seeds coagulate debris in water due to their active soluble protein component, which is a natural cationic polyelectrolyte [84]. Initial water hardness of 80.3 g L^{-1} $CaCO_3$ was found to have decreased between 50% and 70% after coagulation and softening with *M. oleifera* [85]. Water hardness reduction increased with increasing dosage of *M. oleifera* in England using water samples from four sources, with different levels of hardness [86]. Several studies have reported on the performance of *M. oleifera* seeds as an alternative coagulant, or coagulant aid, for various water treatments, such as the removal of turbidity, alkalinity, dissolved organic carbon (DOC), humic acid and hardness from raw water [87–92]. Earlier studies also recommended the use of moringa seed extracts as a coagulant in water treatment for the removal of various pollutants such as acid orange 7 dye and alizarin violet 3R dye [89,93]. Prasad [94] carried out color reduction studies on distillery spent wash using moringa seeds and the optimum color reduction was found to be between 56% and 67% using NaCl and KCl salts respectively. It has also been used to treat palm oil mill effluent waste (POME) and dairy industry waste (DIW) [89,95].

Studies conducted mainly on river water in African countries, including Nigeria, Rwanda, Malawi, Egypt and Sudan, indicated reductions in turbidity and color of over 90% and microorganism (such as *Escherichia coli*) reduction of over 95% [81,84,96]. Moringa seed aqueous extract reduced the number of fecal coliforms, *Staphylococcus aureus*, in water from rivers and wells [96,97]. Moringa seed powder had bacterial removal efficiency of up to 99.5% [96]. In addition, moringa seed powder can reduce heavy metals such as manganese, iron, copper, chromium and zinc in water [98]. The results of several studies conducted in Sudan and England that showed total hardness removal in water treated with moringa seed powder [81]. Water pH is an important parameter that determines coagulating capacity, because it affects the degree of ionization and solubility of adsorbate [99]. The coagulation process using moringa seed powder worked better at pH 6.5–9, while alkaline conditions were better for clarification [84,96,100]. At the same time, high temperatures increased the coagulating power of moringa seed powder and it was negatively affected by temperatures below 15 °C [84]. Therefore, for the best results using *Moringa oleifera* seed powder or extract for water treatment, pH, contact time and temperature should be controlled and monitored [96,101,102]. Overall, moringa is an underused plant that could bring a multifaceted approach to addressing water and nutrition issues in agriculture, particularly in rural sub-Saharan African communities [96].

Moringa seeds can also be used for the purification of water [103]. Aruna and Srilatha [104] reported on the antibacterial effect of moringa seed powder in water purification and clarification of fish ponds. Egbuikwem and Sangodoyin [105] observed 90% turbidity removal by moringa seed extract in well, stream, water samples and examined its effectiveness against *Escherichia coli* in stream water. The presence of 1% flocculent proteins in its oil cakes binds mineral particles and organics in the purification and treatment of drinking water. Moringa seeds present a potential substitute for some conventional synthetic chemical coagulants such as alum—although it is not as effective for turbidity removal as alum—that might increase the risk of cancer and it is considered to be a natural, biodegradable, environmentally friendly and safer substance [104–106].

3.4. Effect of Moringa on Heavy Metal Accumulation in Water and Soil

Heavy metals are some of the most important pollutants affecting water and soil quality. They have significant toxic effects on humans and aquatic species, hence, their removal is essential [89]. Metal biosorption occurs through various mechanisms such as chemisorption, complexation, ion exchange, microprecipitation and adsorption-surface complexation. Metal adsorption onto agro-based adsorbent surfaces occurs owing to the functional groups present in the cell walls of plants [89]. Cellulose, present in the secondary cell wall, adsorbs metals from solution and the metal ions bind either because of two hydroxyl groups present in the cellulose/lignin unit or because of the hydrogen bonds of the metal [89,107]. Metals that can be removed from water using moringa seeds include arsenic, cadmium, zinc and nickel [108–110]. *Moringa oleifera* seeds have been shown to remove arsenic from water [111]. Sorption studies by Idris et al. [89], showed that, in a batch experiment, the optimum conditions achieved for the removal of arsenic (III) and arsenic (V) by *M. oleifera* seeds were 60.21% and 85.06% respectively. For cadmium and nickel, 85.10% and 90% removal were achieved, respectively. The percentage removal was 90% for copper, 80% for lead, 60% for cadmium and 50% for zinc and chromium [89,112].

Moringa seed extract (MSE) removes heavy metal ions from the soil at pH levels of 6 to 8. The success of MSE in removing these elements is owing to its ability to contain them on proteins, creating complexes and limiting their availability [113].

Many studies have shown that the use of MSE has led to a decrease in heavy metal contamination of groundwater [96,114,115]. The use of moringa seeds resulted in the elimination of some heavy metals, such as Fe, while Cu and Cd levels were reduced by 98% and Pb was reduced by 78% [116]. The decrease in heavy metal accumulation in plants and soil can be attributed to the presence of multiple functional groups in MSE, such as carbohydrates, lignin, fatty acids and protein units, which contain a carboxylic group and various amino (-NH$_2$) groups [117] these groups form insoluble complexes with heavy metals, which reduces their availability in the soil to be absorbed by plants [117–119]. MSE is also classified as one of the adsorbents of lignocellulosic, consisting mainly of cellulose, hemicellulose and lignin groups. These groups contain many molecules that can absorb metal ions through ion exchange and complex formation [117].

3.5. Uses of Moringa in Plant Disease Management

Moringa is a multipurpose tree and, recently, a lot of research has gone into its medicinal use against human pathogens [38]. However, Adline and Devi [16] concluded that not as much research has been conducted on the use of moringa as a natural bio-agent against devastating crop pathogens with economic importance. In fact, there is an urgent need to find alternative methods and strategies to help in the management of soil-borne plant diseases; such methods should be ecofriendly and cost-effective. *Moringa oleifera* has been reported to have antimicrobial properties [15] and, therefore, should be included in IPM strategies.

There is no doubt that crops in third world countries, particularly African countries, are victims to fungal toxins (known as mycotoxins), heavy metals and chemicals from the indiscriminate use of pesticides, which present a threat to ecosystems and future generations. These toxins not only lead to imbalances in ecosystems but also interfere with food chains, resulting in environmental abnormalities. It is, therefore, important to find better alternatives for botanical fungicides, which have minimal environmental impact and no danger to human consumption compared with synthetic pesticides [47]. Intensive and continued fungicide usage is associated with development of resistance by fungi to systemic fungicides and the specificity of fungicide formulations, which affect only one pathway in the biosynthesis of fungal pathogens [120], processes that reduce the efficacy of fungicides. Hence, use of biological agents, such as moringa, in the control of soil-borne fungal pathogens notorious for causing root-rot diseases, might prove to be more effective than fungicides [121]. Not only are these bio-agents environmentally friendly compared to chemical methods, but they have also been shown (in several in-vitro studies) to effectively inhibit pathogen growth [122].

Plant extracts from several higher plants have been reported to display antifungal, antibacterial and insecticidal properties under laboratory conditions [123]. This has inspired scientists to investigate large numbers of plants for their antifungal potential against the most important seed-borne fungal species, with the aim of developing some plant-based formulations for the management of plant diseases and maintaining the quality of seeds for sowing and storage [124]. Most of these plant products are reported to have insecticidal properties [125]. It has been reported that the damping-off disease caused by *Pythium debaryanum* can be prevented in seedlings by digging moringa leaves into the soil before planting [22]. Saavedra Gonzalez et al. [126] reported that plants sprayed with moringa leaf ethanol extract were firmer and more resistant to disease and pests. Moreover, moringa has the potential to be used as a biopesticide. Incorporating leaves into the soil can prevent seedling damping off [22]. Furthermore, plants treated with MLE showed greater pest and disease resistance [30].

Foliar application of MLE has been shown to reduce fruit drop and increase fruit set, yield, fruit color, weight, firmness, vitamin C, soluble solids content, anthocyanin content and antioxidant activity in Hollywood plum [127]. Consequently, it may be concluded that the foliar application of moringa leaf aqueous extract can be regarded as a cheap biostimulant—a source of plant growth hormones and minerals—for improving the yield and quality of plant crops, especially given the trend toward organic farming [127]. MLE increased seed germination by 92% compared with a control group and provided seeds with protection from infection: MLE-treated seeds showed 9.33% seed infection, which was significantly different from the control group (66% seed infection) but was not significantly different from chemically treated seeds [124].

Interestingly, in a study of the mycelial growth inhibition of *Aspergillus flavus* isolated from stored maize grains, *M. oleifera* seed powder was the most favorable treatment, compared with Fernazzan D (a chemical fungicidal material) [128].

Several trials have tested the efficacy of MLE in combination with biocontrol agents for plant disease control. In in-vitro experiments, potato dextrose agar (PDA) was amended with MLE and the mycelial growth of *Sclerotium rolfsii* was measured. Results showed that MLE was effective against *Sclerotium* mycelial growth on PDA. Higher aqueous extract concentrations resulted in a decrease in the mycelial growth and no mycelial growth was recorded at an extract concentration [10]. These results show that moringa treatment affected both the mycelial growth of the pathogen and its further development, confirming the antifungal properties of MLE against fungal pathogens [129]. In addition, the combined effect of MLE and other biocontrol agents, revealing that they can be successfully used as a seed treatment against *Sclerotium rolfsii*, the causal agent of damping off and stem rot in cowpea [130].

Moringa is resistant to diseases and pests itself because of its relatively fast vegetative growth, which allows it to regenerate quickly after any disturbance from the most common pests and diseases, including grasshoppers, crickets, caterpillars, termites and fungal disease [126]. A phytochemical analysis of moringa leaves and seed solvent aqueous extracts found flavonoids, alkaloids, tannins, glycosides, terpenoids and phenolic compounds, etc. [131]. *Moringa oleifera* leaves contain some fatty acids, crystalline alkaloids, proteins, niazirin and glycosides, which are thought to be responsible for their antimicrobial activities [129].

The synergistic effects of *Trichoderma* and moringa can protect plant growth against pathogen infection [10]. Farmers could use this combination to decrease the yield losses caused by disease pathogens and therefore increase their income. Microbial diseases are widespread and there is a need to use antimicrobial agents. It has been shown that moringa is an effective antimicrobial agent [132]. MLE can act against bacteria like *Bacillus subtilis*, *Staphylococcus aureus* and *Vibrio cholera* as shown by Viera et al. [133]. The antibacterial effects of the seeds were determined by screening for the antibacterial compounds moringine, benzyl isothiocyanate and pterygospermin [134]. Furthermore, moringa seed aqueous extracts are reported to have antimicrobial properties that can lead to the inhibition of bacterial growth, thereby averting waterborne diseases. These properties of *M. oleifera* seeds not only have a broad application in preventing diseases but can also potentially enhance the quality of life in rural communities [134].

Moreover, moringa can be used in the management of crop disease in organic farming systems due to the various bioactive ingredients it contains, which act in different ways against the pathogenic infection of plants [135]. Farooq et al. [130] conducted an in-vitro study to test the effect of two different concentrations of *Moringa oleifera* leaf and seed aqueous extracts in inhibiting the mycelial growth of two soil-borne pathogens: *Fusarium solani* and *Rhizoctonia solani*. Results showed that both moringa aqueous extracts revealed 50% growth inhibition of *F. solani* at the 30% concentration level. The maximum inhibition percentages recorded against *R. solani* were 45% and 50% using moringa seed aqueous extract at 25% and 30% concentrations, respectively. They concluded that moringa seed and leaf aqueous extracts contain antifungal properties, which resulted in effective growth inhibition of *F. solani* and *R. solani*. Moringa aqueous extract concentration influenced the antifungal activity: the higher concentration levels displayed an increase in antifungal efficacy. Moringa has been used to control the *Rhizopus* pathogen, which is a major causal organism in food spoilage and losses [136]. Many in-vitro studies have highlighted moringa's ability to inhibit food disease pathogens, such as *Salmonella typhi*, *Salmonella paratyphi*, *Escherichia coli*, *Shigella dysenteriae*, *Citrobacter* spp. and *Pseudomonas aeruginosa* [137].

4. Uses of Moringa in the Animal and Poultry Industry

Almost all parts of *Moringa oleifera* are used as food. Leaves of *M. oleifera* are used as food or animal feed during the dry season or periods of drought [138]. In Africa and Asia, moringa pods, flowers, roots and leaves are cooked and eaten as an alternative to green vegetables [13]. The high protein content, good mineral profile and presence of vitamins (especially A, B and C) in moringa leaves make them a feed for animal and poultry. They contain 30% to 40% edible oil (ben oil) [39]. Ben oil provides good amounts of oleic acid, sterols and tocopherols, which prevent rancidity [138] and it possesses antiviral, antioxidant, anti-inflammatory, cardio-protective, anti-asthmatic and anticancer medicinal properties. Antibiotic and antifungal effects against *Fusarium solani*, *Bacillus subtilis*, *Staphylococcus aureus* and *Pseudomonas aeruginosa* result from pterygospermin, which is present in moringa seeds [18,139]. Anemic patients are treated with its leaves to increase their iron levels and its roots and bark are used in the treatment of cardiac issues [103]. According to Jabeen et al. [139] the efficiency of animal feed concentrates can be improved by supplementation with *M. oleifera* leaves. Similarly, Sultana et al. [18] reported that soybean meal (SBM) supplemented with moringa leaves significantly affected growth performance (body weight and body weight gain) in poultry. The birds also had better health status and feed conversion ratio (FCR). According to Chollom et al. [140] MLE has an antiviral effect on Newcastle disease virus (NDV). Despite the great advantages of *Moringa oleifera* when used in animals or poultry, excessive use in large quantities lead to negative results because it contains some anti-nutritional properties that show their effectiveness with the addition of a larger quantity of the animal.

4.1. Inclusion of Moringa oleifera Leaf in Poultry Diets

Moringa oleifera leaves are reportedly devoid of heavy metals such as cadmium, arsenic and mercury and contain significant quantities of vitamins (A, B and C), therefore its integration into poultry diets is safe and could enhance the output performance in poultry production [141]. Adequate levels of dietary moringa leaves could have significant effects on growth, production performance and carcass characteristics of birds. It has been reported that better feed efficiency could be a result of improved digestibility and antimicrobial properties against gut pathogens [142]. Moringa leaf meal can be safely included in cassava-based layer diets at 10% concentration without lowering feed intake. According to Olugbemi et al. [143], *M. oleifera* has a hypocholesterolemic effect and it can be added to poultry feed to reduce egg cholesterol content. Abou-Elezz et al. [144] verified that dietary moringa leaf meal (up to 10%) can produce a useful impact on yolk color and resulted in no significant adverse effects on egg laying rate. Therefore, 10% moringa leaf meal has been recommended as a sustainable feed supplement in laying hen diets. The 5% moringa leaf meal level had a beneficial impact on birds, while dietary levels of 15% and 20% produced adverse effects [144,145]. Safa and Tazi [146] verified that broilers fed diets containing 5% moringa leaf meal for 7 weeks had increased body weight and higher total feed intake and improved feed conversion ratio compared to a control group. In another experiment, dry matter intake (DMI) increased as moringa leaf meal inclusion in feed increased in broilers because of increased bulk and metabolizable concentration [147]. Moringa leaf meal was a good source for improving yolk pigments and had no negative effects on egg shape index and shell thickness [148]. This could be due to the high carotene content (~15.25 to 16.30 mg 100 g^{-1} of moringa leaf meal) [149]. Higher albumen and lower yolk, indices imply a relatively lower concentration of cholesterol, which is a high-quality attribute for egg consumers [148]. The supplementation of sunflower with moringa leaf meal had a significant effect on egg weight when used at 5% concentration in the diet [145].

Moringa oleifera has antioxidant activity due to its phytochemical content, which can influence the stability, palatability, processing properties and shelf life of poultry products [43,150]. Flavonoids, mainly flavonols, are the most important antioxidants in moringa [39]. Their antioxidant activity was higher than that of vitamin C and could be used to prolong the shelf life of poultry products [151]. Dietary supplementation with low levels of moringa leaf meal has no effect on nutrient digestibility and may even improve feed efficiency. Supplementary moringa leaf meal of up to 10% showed no significant effects on body weight, feed consumption, or feed conversion ratio [150]. The positive effects of dietary *Moringa oleifera* supplementation on animal performance may be attributed to contents of moringa in calcium, magnesium, sodium, potassium, copper, iron, zinc, manganese, α-tocopherol, β-carotene and ascorbic acid as well as PUFA and some bioactive components of moringa.

Other studies have also established that excessive amounts of *M. oleifera* might produce adverse effects on egg-laying performance. Addition of 20% moringa leaf meal to layer diets, as a substitute for sunflower seed meal, significantly decreased egg production and total egg weight [145]. Similarly, Mutayoba et al. [152] reported that supplementing diets with 20% moringa leaf meal adversely affected egg mass production and egg laying rate, despite the dietary energy level, while the 5% supplementation level had no adverse effects. Ebenebe et al. [153] observed that diets supplemented with 2.5% moringa leaf meal significantly affected internal egg quality in comparison with the control group; adding varying grades of moringa leaf meal (0%, 5%, 10% and 15%) to layer diets linearly decreased egg mass and egg-laying percentage, while egg weight showed a quadratic trend as the percentage of leaf meal increased [143]. The possibility of moringa leaf meal being used as a substitute for soybean meal in broiler feed, but reported that high levels of leaf meal led to a decrease in growth rate [149]. According to Olugbemi et al. [143], these adverse effects of high levels of leaf meal in poultry diets could be a result of low digestibility of the protein. This conclusion is supported by

Kakengi et al. [145], who observed an increase in feed intake and dry matter intake in laying hens fed diets containing 10% and 20% moringa leaf meal. This result was similar to previous reports [144] that dry matter intake showed a quadratic trend with increasing levels moringa leaf meal (0–15%). However, feed efficiency was negatively affected in the birds fed diets supplemented with higher levels of leaf meal [145]. Moringa leaf meal could be used to replace the groundnut cake in the diets of grower rabbits and the supplementation level can reach 60% with high feed to gain ratio and feed cost efficiency [154]. Phytate content in M. olifera might decrease availability of minerals, thus various enzymes such as phytase could be added to feed containing moringa leaves to increase phosphorus availability [55,155].

Several investigations exhibited that poultry performance was depressed with addition of *M. oleifera*, evidenced by decreased body gain and increased feed conversion ratio, which might be due to the anti-nutritional factors, such as phytate tannins, total phenols and saponins [30]. Thus, more work is required to examine how different levels of supplementary *M. oleifera* affect the poultry performance and further limit an optimal inclusion rate in diets. In general, it means the plant can be dangerous only if consumed too frequently or in large amounts. Supplementation with *M. oleifera* at low levels improved egg quality but higher levels of inclusion resulted in lower productivity [144]. Decreased egg weight at higher concentrations of moringa leaf meal ($\geq$10%) was due to lower protein retention, energy availability and lower crude fiber (CF) digestibility [156]. These observations suggested that *M. oleifera* leaves might be a suitable feed stuff for poultry; however, attention should be paid to the dietary levels.

4.2. Anticoccidial Effect of Moringa oleifera on Poultry Parasitic Diseases

One of the most important diseases of poultry is avian coccidiosis and it is responsible for many broiler mortalities worldwide. The use of anticoccidial drugs is the main control of this disease; however, herbal preparations could be used as a replacement coccidiosis treatment in chickens. Ola-Fadunsin et al. [157] examined the efficacy of *M. oleifera* acetone extracts (1.0 to 5.0 g/kg body weight) against avian coccidiosis and found a direct effect on broiler chickens infected naturally with mixed *Eimeria* species compared with negative control (untreated group) and positive control (treated with toltrazuril, 7 mg/kg BW). The anticoccidial activity of *M. oleifera*, administered either as powdered leaves, as a prophylactic, or as ethanolic extract, could be related to the antioxidant properties of *M. oleifera* (ascorbic acid, flavonoids, phenolics and carotenoids) [158]. These compounds inhibit the presence of oocysts in fecal matter, provide cellular protection against oxidative stress and decrease the severity of *E. tenella* infections by changing the degree of peroxidation of the intestinal lipid [159].

Allen et al. [159] examined the antioxidant effect of *M. oleifera* leaf ethanol extract and fruit. After infection with *Eimeria* species, the host's cellular immune response produced free radical oxidative species, which play a vital role in the defense mechanism against parasitic infections. [160] observed that antioxidant activity is due to the presence of polyphenols, tannins, anthocyanin, glycosides and thiocarbamates, which may remove free radicals, activate antioxidant enzymes and inhibit oxidases because of cytoplasmic membrane attachment and, thus, make these elements available for the birds to use [139].

4.3. Antiviral Activity of Moringa oleifera on Poultry Viral Disease

Newcastle disease virus is considered one of the most infectious and contagious viral diseases of domestic poultry and wild birds. It has a high morbidity and mortality rate, which can result in sharp economic losses to the poultry industry worldwide [161]. The main control system of NDV is vaccination but there are some challenges with this approach, especially in rural areas with poor farms: the cost of vaccines is high, the cold chain systems required by these vaccines may not be available and small size and multiage birds, can affect the success of the vaccination [162]. Improving the immunogenicity of the vaccine by applying complementary approaches, such as natural plants, might be a good way to overcome such infectious diseases.

A complementary method of controlling this virus is the use of medicinal plants. Plants contain alkaloids, flavonoids, saponins and tannins, which can act as antiviral agents. Several studies have examined the moringa extract concentration required to provide the best antiviral activity. The effect of aqueous seed aqueous extract of *M. oleifera* against NDV was investigated by Chollom et al. [140], who used an in ovo assay and reported that the extract concentration was directly proportional to virus death and inversely proportional to the production of antibodies against NDV. According to these findings, *M. oleifera* seed aqueous extract has a powerful antiviral activity against NDV in ovo; it also had a nutritional value. The extract contains considerable amounts of vitamins A, B and C, minerals (such as calcium ions, iron, potassium) and proteins, in addition to traces of carotenoids, saponins, phytates and phenolic constituents [42,163] which may be responsible for the immunomodulation of the immune system. The role of *M. oleifera* in modulating immune responses may be linked to the enhanced production of factors responsible for growth, such as cytokines, which activate both innate and adaptive immunity [164].

4.4. Antibacterial Effect of Moringa oleifera on Poultry Bacterial Disease

Pathogens that cause disease and economic losses in poultry include *Escherichia coli*, *Salmonella* spp. and mycoplasma [165]. Abiodun et al. [166] studied the antibacterial and phytochemical effects of aqueous extracts of *M. oleifera* roots on *Escherichia coli*-infected broiler chicks and established that moringa roots (aqueous extract) can be used as a replacement for synthetic antibiotics in combating pertinent poultry diseases, especially those of the *Escherichia coli* origin. Extracts of 15 g L^{-1} dosage are recommended, since this level shows better serological indices than other dose levels examined (5 or 10 g L^{-1}) compared with commercial antibiotics [166]. *Moringa oleifera* acetone extract was reported to have antibacterial properties and conclusion was made to investigate it as a phytotherapeutic agent to combat infectious agents [167]. The antimicrobial action of *M. oleifera* seed extracts might be due to the presence of lipophilic compounds; these compounds can attach to the cytoplasmic membrane. The moringa seed ethanol extract may also contain antibiotic metabolites such as carboxylic acid, 2,4-diacetylphloroglucinol, cell wall-degrading enzymes and chitinases [136,139]. From the previous research literature that was interested in the antibacterial effect of *Moringa oleifera* on poultry we found it rare, so we encourage more studies at this point.

5. Conclusions

Moringa oleifera L. is known as one of the most useful multipurpose plants. Some parts of the moringa tree (leaves, pods, seeds, flowers, fruits and roots) are eaten as food and some are taken as a remedy. Moringa is rich in phytochemical compounds that confer on the plant significant medicinal properties that could be valuable for treating certain ailments. The leaves and seeds of *M. oleifera* are a source of protein, iron, calcium, ascorbic acid vitamin A and antioxidant compounds such as carotenoids, flavonoids, vitamin E and phenolic compounds. Thus, its leaves could be used as a supplement to improve feed efficiency and livestock performance, or be used to replace conventional crops to obtain more economically sustainable. The use of moringa as a crop enhancer is an environmentally friendly strategy for improving crop yields at the lowest possible cost. In addition, moringa and its derivatives have several nutritional and biological applications, including use in green fertilization, animal and poultry feeds, medicines, biopesticides and seed production. In addition, moringa contributes in plant disease management in being antioxidant, antifungal, antibacterial and insecticidal. On the other hand, some parts of the moringa tree contain toxins and other anti-nutritional factors, which limit their utility as a source of food for humans or animals. This means that the plant can be dangerous if consumed too frequently or in large amounts. Therefore, it has been suggested that more attention should be paid to how moringa is used in diets. We recommend the use of moringa at low or medium concentrations in the field of animal and poultry production because high concentrations cause some problems because it contains some toxins and other anti-nutritional factors. However, the benefit of moringa should be used to improve plant production and soil characteristics.

Therefore, emphasis should be placed on providing research on moringa to highlight its roles in agricultural production (plant and animal).

Author Contributions: Conceptualization, M.E.A.E.-H., M.A., A.S.E., E-S.M.D., H.M.N.T. and A.S.M.E.; Writing-Original Draft Preparation, M.E.A.E.-H., M.A., A.S.E., E-S.M.D., H.M.N.T., A.S.M.E, S.S.E. and A.A.S; Writing-Review & Editing, M.E.A.E.-H., M.A., A.S.E., E-S.M.D., H.M.N.T. and A.S.M.E.; Visualization, M.E.A.E.-H. and M.A.; Administration, S.S.E. and A.A.S.

Funding: This research received no external funding.

Acknowledgments: Authors thank their respective universities for their support.

Conflicts of Interest: The authors declare no conflict of interest.

References

1. Ahossi, P.K.; Dougnon, J.T.; Kiki, P.S.; Houessionon, J.M. Effects of *Tridax procumbens* powder on zootechnical, biochemical parameters and carcass characteristics of Hubbard broiler chicken. *J. Anim. Health Prod.* **2016**, *4*, 15–21.

2. Ciftci, M.; Ertas, O.N.; Guler, T. Effect of vitamin E and vitamin C dietary supplementation on egg production and egg quality of laying hens exposed to a chronic stress. *Rev. Med. Vet.* **2005**, *156*, 107–111.

3. Nghonjuyi, N.W.; Tiambo, C.K.; Kimbi, H.K.; Manka'a, C.N.; Juliano, R.S.; Lisita, F. Efficacy of ethanolic extract of Carica papaya leaves as a substitute of sulphanomide for the control of coccidiosis in KABIR chickens in Cameroon. *J. Anim. Health Prod.* **2015**, *3*, 21–27. [CrossRef]

4. Cross, D.E.; Mcdevith, R.M.; Hillman, K.; Agamovic, T. The effect of herbs and their associated essential oils on performance, digestibilities and gut microflora in chickens 7 to 28 days of age. *Br. Poult. Sci.* **2007**, *4*, 496–506. [CrossRef] [PubMed]

5. Soliman, K.M.; Badea, R.I. Effect of oil extracted from some medicinal plants on different mycotoxigenic fungi. *Food Chem. Toxicol.* **2002**, *40*, 1669–1675. [CrossRef]

6. Burtis, M.; Bucar, F. Antioxidant activity of *Nigella sativa* essential oil. *Phytother. Res.* **2000**, *14*, 323–328. [CrossRef]

7. Abd El-Hack, M.E.; Alagawany, M. Performance, egg quality, blood profile, immune function and antioxidant enzyme activities in laying hens fed diets with thyme powder. *J. Anim. Feed Sci.* **2015**, *24*, 127–133. [CrossRef]

8. Abd El-Hack, M.E.; Alagawany, M.; Farag, M.R.; Tiwari, R.; Kumaragurubaran, K.; Dhama, K.; Zorriehzahra, J.; Adel, M. Beneficial Impacts of Thymol Essential Oil on Health and Production of Animals, Fish and Poultry: A review. *J. Essent. Oil Res.* **2016**, *28*. [CrossRef]

9. Maroni, H.M. Studying the effects of pesticides on humans (International Commission on Occupational Health). In Proceedings of the 9th International workshop, International Centre for Pesticide Safety, Busto Garolfo, Milan, Italy, 2–4 May 1990.

10. Adandonon, A.; Aveling, T.A.S.; Labuschagne, N.; Tamo, M. Biocontrol agents in combination with *Moringa oleifera* extract for integrated control of *Sclerotium*-caused cowpea damping-off and stem rot. *Eur. J. Plant Pathol.* **2006**, *115*, 409–418. [CrossRef]

11. Farag, M.R.; Alagawany, M.; Badr, M.M.; Khalil, S.R.; El-Kholy, M.S. An overview of *Jatropha curcas* meal induced productive and reproductive toxicity in Japanese quail: Potential mechanisms and heat detoxification. *Theriogenology* **2018**, *113*, 208–220. [CrossRef] [PubMed]

12. Dahot, M.U. Vitamin contents of the flowers and seeds of *Moringa oleifera. Pak. J. Biochem.* **1988**, *21*, 1–2.

13. Fahey, J.W.; Zalcmann, A.T.; Talalay, P. The chemical diversity and distribution of glucosinolates and isothiocyanates among plants. *Phytochemistry* **2001**, *56*, 5–51. [CrossRef]

14. Yang, R.; Chang, L.C.; Hsu, J.C.; Weng, B.B.C.; Palada, M.C.; Chadha, M.L.; Levasseur, V. Nutritional and functional properties of Moringa leaves—from Germplasm, to plant, to food, to health. *Moringa Nutr. Plant Resour. Strateg. Stand. Mark. Better Impact Nutr. Afr.* **2006**, *11*, 16–18.

15. Ali, G.H.; El-Taweel, G.E.; Ali, M.A. The cytotoxicity and antimicrobial efficiency of *Moringa oleifera* seeds extracts. *Int. J. Environ. Stud.* **2004**, *61*, 699–708. [CrossRef]

16. Adline, J.; Devi, J. A study on phytochemical screening and antibacterial activity of *Moringa oleifera. Int. J. Res. Appl.* **2014**, *2*, 169–176.

17. Fuglie, L.J. *The Miracle Tree: Moringa oleifera, Natural Nutrition for the Tropics*; Church World Service: Elkhart, IN, USA, 1999; p. 172.

18. Sultana, N.; Alimon, A.R.; Haque, K.S.; Sazili, A.Q.; Yaakub, H.; Hossain, S.M.J. The effect of cutting interval on yield and nutrient composition of different plant fractions of Moringa oleifera tree. *J. Food Agric. Environ.* **2014**, *12*, 599–604.

19. Olson, M.E.; Carlquist, S. Stem and root anatomical correlations with life form diversity, ecology and systematics in *Moringa* (Moringaceae). *Bot. J. Linn. Soc.* **2001**, *135*, 315–348. [CrossRef]

20. Olson, M.E. Combining data from DNA sequences and morphology for a phylogeny of Moringaceae (Brassicales). *Syst. Bot.* **2002**, *27*, 55–73.

21. Shahzad, U.; Khan, M.A.; Jaskani, M.J.; Khan, I.A.; Korban, S.S. Genetic diversity and population structure of *Moringa oleifera*. *Conserv. Genet.* **2013**, *14*, 1161–1172. [CrossRef]

22. Fahey, J.W. *Moringa oleifera*: A Review of the Medical Evidence for Its Nutritional, Therapeutic and Prophylactic Properties. *Trees Life J.* **2005**, *1*, 1–15.

23. Berger, M.R.; Habs, M.; Jahn, S.A.A.; Schmahl, D. Toxicological assessment of seeds from *Moringa oleifera* and *Moringa stenopetala*, two highly efficient primary coagulants for domestic water treatment of tropical waters. *East Afr. Med. J.* **1984**, *61*, 712–717. [PubMed]

24. Sultana, B.; Anwar, F. Flavonols (kaempeferol, quercetin, myricetin) contents of selected fruits, vegetables and medicinal plants. *Food Chem.* **2008**, *108*, 879–884. [CrossRef] [PubMed]

25. Gopalakrishnan, L.; Doriya, K.; Kumar, D.S. *Moringa oleifera*: A review on nutritive importance and its medicinal application. *Food Sci. Hum. Wellness* **2016**, *5*, 49–56. [CrossRef]

26. Aregheore, E.M. Intake and digestibility of *Moringa oleifera*-batiki grass mixtures for growing goats. *Small Rumin. Res.* **2002**, *46*, 23–28. [CrossRef]

27. Richter, N.; Perumal, S.; Becker, K. Evaluation of nutritional quality of Moringa (*Moringa oleifera* Lam.) leaves as an alternative protein source for Nile tilapia (*Oreochromis niloticus* L.). *Aquaculture* **2003**, *217*, 599–611. [CrossRef]

28. Rubanza, C.; Shem, M.; Otsyina, E.; Bakengesa, S.; Ichinohe, T.; Fujihara, T. Polyphenolics and tannins effect on in vitro digestibility of selected Acacia species leaves. *Anim. Feed Sci. Technol.* **2005**, *119*, 129–142. [CrossRef]

29. Asaolou, V.; Binuomote, R.; Akinlade, J.; Aderinola, O.; Oyelami, O. Intake and growth performance of West African dwarf goats fed *Moringa oleifera*, *Gliricidia Sepium* and *Leucaena leucocephala* dried leaves as supplements to cassava peels. *J. Biol. Agric. Healthc.* **2012**, *2*, 76–88.

30. Makkar, H.P.S.; Becker, K. Nutritional value and antinutritional components of whole and ethanol extracted of *Moringa oleifera* leaves. *Anim. Feed Sci. Technol.* **1996**, *63*, 211–228. [CrossRef]

31. Grubben, G.; Denton, O. *Plant Resources of Tropical Africa*; Earthprint Limited: Stevenage, UK, 2004; pp. 56–60.

32. Bose, C.K. Nerve growth factor, follicle stimulating hormone receptor and epithelial ovarian cancer. *Med. Hypotheses* **2004**, *5*, 917–918. [CrossRef] [PubMed]

33. Bhattacharya, J.; Guha, G.; Bhattacharya, B. Powder microscopy of bark–poison used for abortion: *Moringa pterygosperma* gaertn. *J. Indian Forensic Sci.* **1978**, *17*, 47–50. [PubMed]

34. Faizi, S.; Siddiqui, B.S.; Saleem, R.; Siddiqui, S.; Aftab, K.; Gilani, A.H. Isolation and structure elucidation of new nitrile and mustard oil glycosides from *Moringa oleifera* and their effect on blood pressure. *J. Nat. Prod.* **1994**, *57*, 1256–1261. [CrossRef] [PubMed]

35. Morton, J.F. The Horseradish Tree, *Moringa Pterygosperma* (Moringaceae)—A boon to arid lands? *Econ. Bot.* **1991**, *45*, 318–333. [CrossRef]

36. Mazumder, U.K.; Gupta, M.; Chakrabarti, S.; Pal, D. Evaluation of hematological and hepatorenal functions of methanolic extract of *Moringa oleifera* Lam root treated mice. *Indian J. Exp. Biol.* **1999**, *37*, 612–614. [PubMed]

37. Bennett, R.; Mellon, F.; Foidl, N.; Pratt, J.; Dupont, M.; Perkins, L.; Kroon, P. Profiling glucosinolates and phenolics in vegetative and reproductive tissues of the multi-purpose trees *Moringa oleifera* L. (Horseradish tree) and *Moringa stenopetala* L. *J. Agric. Food Chem.* **2003**, *51*, 3546–3553. [CrossRef] [PubMed]

38. Anwar, F.; Latif, S.; Ashraf, M.; Gilani, A.H. *Moringa oleifera*: A food plant with multiple medicinal uses. *Phytother. Res.* **2007**, *21*, 17–25. [CrossRef] [PubMed]

39. Pandey, A. *Moringa oleifera* Lam. (Sahijan)—A plant with a plethora of diverse therapeutic benefits: An updated retrospection. *Med. Aromat. Plants* **2012**, *1*, 101. [CrossRef]

40. Al-Asmari, A.K.; Albalawi, S.M.; Athar, M.T.; Khan, A.Q.; Al-Shahrani, H.; Islam, M. *Moringa oleifera* as an anti-cancer agent against breast and colorectal cancer cell lines. *PLoS ONE* **2015**, *10*. [CrossRef] [PubMed]

41. Yameogo, C.; Bengaly, M.; Savadogo, A.; Nikiema, P.; Traore, S. Determination of chemical composition and nutritional values *Moringa oleifera* leaves. *Pak. J. Nutr.* **2011**, *10*, 264–268. [CrossRef]

42. Siddhuraju, P.; Becker, K. Antioxidant properties of various solvent extracts of total phenolic constituents from three different agro-climatic origins of drumstick tree (*Moringa oleifera* Lam.). *J. Agric. Food Chem.* **2003**, *15*, 2144–2155. [CrossRef] [PubMed]

43. Abbas, T.; Ahmed, M. Use of *Moringa oleifera* seeds in broilers diet and its effects on the performance and carcass characteristics. *Int. J. Appl. Poult. Res.* **2012**, *1*, 1–4.

44. Pakade, V.; Cukrowska, E.; Chimuka, L. Comparison of antioxidant activity of *Moringa oleifera* and selected vegetables in South Africa. *S. Afr. J. Sci.* **2013**, *109*, 3–4. [CrossRef]

45. Anwar, F.; Bhanger, M. Analytical characterization of *Moringa oleifera* seed oil grown in temperate regions of Pakistan. *J. Agric. Food Chem.* **2003**, *51*, 6558–6563. [CrossRef] [PubMed]

46. Sreelatha, S.; Padma, P. Antioxidant activity and total phenolic content of *Moringa oleifera* leaves in two stages of maturity. *Plant Foods Hum. Nutr.* **2009**, *64*, 303–311. [CrossRef] [PubMed]

47. Verma, J.; Dubey, N.K. Prospectives of botanicals and microbial products as pesticides of tomorrow. *Curr. Sci.* **1999**, *76*, 172–179.

48. Jaafaru, M.S.; Nordin, N.; Shaari, K.; Rosli, R.; Abdull Razis, A.F. Isothiocyanate from *Moringa oleifera* seeds mitigates hydrogen peroxide-induced cytotoxicity and preserved morphological features of human neuronal cells. *PLoS ONE* **2018**, *13*. [CrossRef]

49. Yasmeen, A.; Basra, S.M.A.; Ahmad, R.; Wahid, A. Performance of late sown wheat in response to foliar application of *Moringa oleifera* Lam. leaf extract. *Chil. J. Agric. Res.* **2012**, *2*, 92–97. [CrossRef]

50. Rehman, H.; Nawaz, M.Q.; Basra, S.M.A.; Afzal, I.; Yasmeen, A.; Hassan, F.U. Seed priming influence on early crop growth, phenological development and yield performance of linola (*Linum usitatissimum* L.). *J. Integr. Agric.* **2014**, *13*, 990–996. [CrossRef]

51. Rady, M.M.; Bhavya, V.C.; Howladar, S.M. Common bean (*Phaseolus vulgaris* L.) seedlings overcome NaCl stress as a result of presoaking in *Moringa oleifera* leaf extract. *Sci. Hortic.* **2013**, *162*, 63–70. [CrossRef]

52. Salem, J.M. In vitro propagation of *Moringa oleifera* L. under salinity and ventilation conditions. *Genet. Plant Physiol.* **2016**, *6*, 54–64.

53. Rashid, U.; Anwar, F.; Moser, B.R.; Knothe, G. *Moringa oleifera* oil: A possible source of biodiesel. *Bioresour. Technol.* **2008**, *99*, 8175–8179. [CrossRef] [PubMed]

54. Bashir, K.A.; Waziri, A.F.; Musa, D.D. *Moringa oleifera*, a potential miracle tree; a review. *IOSR J. Pharm. Biol. Sci.* **2016**, *11*, 25–30.

55. Fuglie, L. New uses of Moringa studied in Nicaragua: Moringa leaf concentrate. *ECHO Dev. Notes* **2009**, *68*, 1–8.

56. Howladar, S.M. A novel *Moringa oleifera* leaf extract can mitigate the stress effects of salinity and cadmium in bean (*Phaseolus vulgaris* L.) plants. *Ecotoxicol. Environ. Saf.* **2014**, *100*, 69–75. [CrossRef] [PubMed]

57. Rady, M.M.; Mohamed, G.F.; Abdalla, A.M.; Ahmed Yasmin, H.M. Integrated application of salicylic acid and *Moringa oleifera* leaf extract alleviates the salt-induced adverse effects in common bean plants. *Int. J. Agric. Technol.* **2015**, *11*, 1595–1614.

58. Osman, H.; Abohassan, A.A. *Moringa: The Strategic Tree for the Third Century*; King Abdulaziz University Publishing Center: Jeddah, Saudi Arabia, 2015; ISBN 7-745-06-9960.

59. Amirigbal, M.; Nadeemakbar, A.; Abbas, R.; Khan, H.; Maqsood, Q. Response of Canola to foliar application of Moring (*Moringa oleifera* L.) and Brassica (*Brassica napus* L.) water extracts. *Int. J. Agric. Crop Sci.* **2014**, *14*, 1431–1433.

60. Muhammed, R.; Olurokooba, M.; Akinyaju, J.; Kambai, E. Evaluation of different concentrations and frequency of foliar application of Moringa extract on growth and yield of onions. *Agroresearch* **2013**, *13*, 196–205. [CrossRef]

61. Azooz, M.M.; Shaddad, M.A.; Abdel-Latef, A.A. The accumulation and compartmentation of proline in relation to salt tolerance of three sorghum cultivars. *Ind. J. Plant Physiol.* **2004**, *9*, 1–8.

62. Afzal, I.; Hussain, B.; Basra, S.M.A.; Rehman, H. Priming with MLE reduces imbibitional chilling injury in spring maize. *Seed Sci. Technol.* **2012**, *40*, 271–276. [CrossRef]

63. Basra, S.M.A.; Iftikhar, M.N.; Afzal, I. Potential of moringa (*Moringa oleifera*) leaf extract as priming agent for hybrid maize seeds. *Int. J. Agric. Biol.* **2011**, *13*, 1006–1010.

64. Moyo, B.; Masika, P.J.; Hugo, A.; Muchenje, V. Nutritional characterization of Moringa (*Moringa oleifera* Lam) leaves. *Afr. J. Biotech.* **2011**, *10*, 12925–12933.

65. Taiz, L.; Zeiger, E. *Plant Physiology*; Sinauer Associates: Sunderland, MA, USA, 2010.

66. Desoky, E.M.; Merwad, A.M.; Elrys, A.S. Response of pea plants to natural bio-stimulants under soil salinity stress. *Am. J. Plant Physiol.* **2017**, *12*, 28–37.

67. Rady, M.M.; Mohamed, G.F. Modulation of salt stress effects on the growth, physio-chemical attributes and yields of *Phaseolus vulgaris* L. plants by the combined application of salicylic acid and *Moringa oleifera* leaf extract. *Sci. Hortic.* **2015**, *193*, 105–113. [CrossRef]

68. Basra, S.M.A.; Afzal, I.; Anwar, S.; Shafique, M.; Haq, A.; Majeed, K. Effect of different seed invigoration techniques on wheat (*Triticum aestivum* L.) seeds sown under saline and non-saline conditions. *J. Seed Technol.* **2005**, *28*, 36–45.

69. Nouman, W.; Siddiqui, M.T.; Basra, S.M.A.; Farooq, H.; Zubair, M.; Gull, T. Biomass production and nutritional quality of *Moringa oleifera* as field crop. *Turk. J. Agric. For.* **2013**, *37*, 410–419. [CrossRef]

70. Azra, Y. Exploring the Potential of Moringa (*Moringa Oleifera*) Leaf Extract as a Natural Plant Growth Enhancer. Ph.D. Thesis, University of Agriculture, Faisalabad, Pakistan, 2011.

71. Hanafy, R. Using *Moringa oleifera* leaf extract as a bio-fertilizer for drought stress mitigation of *Glycine max* L. plants. *Egypt. J. Bot.* **2017**, *57*, 281–292.

72. Dietrich, J.T.; Kaminek, V.; Belvins, D.G.; Reinbett, T.M.; Morris, R.D. Changes in cytokinins and cytokinin oxidase activity in developing maize kernel and the effects of exogenous cytokinin on kernel development. *Plant Physiol. Biochem.* **1995**, *33*, 327–336.

73. Stevens, J.; Senaratna, T.; Sivasithamparam, K. Salicylic acid induces salinity tolerance in tomato (*Lycopersicon esculentum* cv. Roma): Associated changes in gas exchange, water relations and membrane stabilization. *Plant Growth Regul.* **2006**, *49*, 77–83.

74. Gonzalez, L.; Gonzalez-Vilar, M. Determination of relative water content. In *Handbook of Plant Ecophysiology Techniques*; Reigosa, M.J., Ed.; Springer: Dordrecht, The Netherlands, 2001; pp. 207–212.

75. Hebers, K.; Sonnewald, V. Altered gene expression: Brought about by inter and pathogen interactions. *J. Plant Res.* **1998**, *111*, 323–328. [CrossRef]

76. Hernández, J.A.; Francisco, F.J.; Corpas, G.M.; Gómez, L.A.; Del Río, F.S. Salt induced oxidative stresses mediated by activated oxygen species in pea leaves mitochondria. *Physiol. Plant.* **1993**, *89*, 103–110. [CrossRef]

77. Del Río, L.A.; Sandalio, L.M.; Corpas, F.J.; Palma, J.M.; Barroso, J.B. Reactive oxygen species and reactive nitrogen species in peroxisomes, production scavenging and role in cell signaling. *Plant Physiol.* **2006**, *141*, 330–335. [CrossRef] [PubMed]

78. Zaki, S.F.; Rady, M.M. *Moringa oleifera* leaf extract improves growth, physio-chemical attributes, antioxidant defence system and yields of salt-stressed *Phaseolus vulgaris* L. plants. *Int. J. Chem. Technol. Res.* **2015**, *8*, 120–134.

79. Foyer, C.H. The role of ascorbate in plants, interactions with photosynthesis and regulatory significance. In *Active Oxygen, Oxidative Stress and Plant Metabolism*; American Society of Plant Physiology: Monona Drive, MD, USA, 1991.

80. Tetley, R.M.; Thimann, K.V. The metabolism of oat leaves during senescence: I. Respiration, carbohydrate metabolism and the action of cytokinins. *Plant Physiol.* **1974**, *54*, 294–303. [CrossRef] [PubMed]

81. Bichi, M.H. A Review of the Applications of *Moringa oleifera* Seeds Extract in Water Treatment. *Civ. Environ. Res.* **2013**, *3*, 1–8.

82. Kansal, S.K.; Kumari, A. Potential of *M. oleifera* for the treatment of water and wastewater. *Chem. Rev.* **2014**, *114*, 4993–5010. [CrossRef] [PubMed]

83. Deeba, F.; Abbas, N.; Butt, T.; Imtiaz, N.; Khan, R.A.; Ahsan, M.M. Utilization of *Moringa oleifera* seeds for treatment of canal and industrial waste water-an alternative sustainable solution for developing countries. *J. Biodivers. Environ. Sci.* **2015**, *7*, 54–60.

84. Pritchard, M.; Craven, T.; Mkandawire, T.; Edmondson, A.; O'neill, J. A study of the parameters affecting the effectiveness of *Moringa oleifera* in drinking water purification. *Phys. Chem. Earth Parts A/B/C* **2010**, *35*, 791–797. [CrossRef]

85. Barth, H.; Habs, M.; Klute, R.; Muller, S.; Bernard, T. Anwendung Von naturlichen Wirkstoffen aus *Moringa oliefera* Lam Samen zur rinkwasseruabereitung. *Chem. Ztg.* **1982**, *1*, 75–78. (In German)

86. Adedapo, A.; Mogbojuri, O.; Emikpe, B. Safety evaluations of the aqueous extract of the leafs of Moringa oleifera in rats. *J. Med. Plants Res.* **2009**, *3*, 586–591.

87. Ali, E.N. Application of Moringa Seeds Extract in Water Treatment. Ph.D. Thesis, International Islamic University, Kuala Lumpur, Malaysia, 2010.

88. Fahmi, M.R.; Najib, N.W.A.Z.; Ping, P.C.; Hamidin, N. Mechanism of turbidity and hardness removal in hard water sources by using *Moringa oleifera*. *J. Appl. Sci.* **2011**, *11*, 2947–2953. [CrossRef]

89. Idris, M.A.; Jami, M.S.; Hammed, A.M.; Jamal, P. *Moringa oleifera* seed extract: A review on its environmental applications. *Int. J. Appl. Environ. Sci.* **2016**, *11*, 1469–1486.

90. Lea, M. Bioremediation of turbid surface water using seed extract from *Moringa oleifera* Lam. (drumstick) tree. *Curr. Protoc. Microbiol.* **2010**, *16*, 1–2.

91. Muyibi, S.A.; Okuofu, C.A. Softening hard well waters with *Moringa oleifera* seed extracts. *Int. J. Environ. Stud.* **1996**, *50*, 247–257. [CrossRef]

92. Santos, A.F.S.; Paiva, P.M.G.; Teixeira, J.C.; Brito, A.G.; Coelho, L.C.B.B.; Nogueira, R. Coagulant properties of *Moringa oleifera* protein preparations: Application to humic acid removal. *Environ. Technol.* **2013**, *33*, 69–75. [CrossRef] [PubMed]

93. Marandi, R.; Bakhtiar Sepehr, S.M. Removal of orange 7 dye from wastewater used by natural adsorbent of *Moringa oleifera* seeds. *Amer. J. Environ. Eng.* **2012**, *1*, 1–9. [CrossRef]

94. Prasad, R.K. Color removal from distillery spent wash through coagulation using *Moringa oleifera* seeds: Use of optimum response surface methodology. *J. Hazard. Mater.* **2009**, *165*, 804–811. [CrossRef] [PubMed]

95. Bhatia, S.; Othman, Z.; Ahmad, A.L. Pretreatment of palm oil mill effluent (POME) using *Moringa oleifera* seeds as natural coagulant. *J. Hazard. Mater.* **2007**, *145*, 120–126. [CrossRef] [PubMed]

96. James, A.; Zikankuba, V. *Moringa oleifera* a potential tree for nutrition security in sub-Sahara Africa. *Am. J. Res. Commun.* **2017**, *5*, 1–12.

97. Ferreira, R.S.; Napoleão, T.H.; Santos, A.F.S.; Sá, R.A.; Carneiro-da-Cunha, M.G.; Morais, M.M.C.; Silva-Lucca, R.A.; Oliva, M.L.V.; Coelho, L.C.B.B.; Paiva, P.M.G. Coagulant and antibacterial activities of the water-soluble seed lectin from *Moringa oleifera*. *Appl. Microbiol.* **2011**, *53*, 186–192. [CrossRef] [PubMed]

98. Nkurunziza, T.; Nduwayezu, J.B.; Banadda, E.N.; Nhapi, I. The effect of turbidity levels and *Moringa oleifera* concentration on the effectiveness of coagulation in water treatment. *Water Sci. Technol.* **2009**, *59*, 1551–1558. [CrossRef] [PubMed]

99. Sivakumar, D. Adsorption study on municipal solid waste leachate using *Moringa oleifera* seed. *Int. J. Environ. Sci. Technol.* **2013**, *10*, 113–124. [CrossRef]

100. Thakur, S.S.; Choubey, S. Assessment of coagulation efficiency of *Moringa oleifera* and Okra for treatment of turbid water. *Arch. Appl. Sci. Res.* **2014**, *6*, 24–30.

101. Alsharaa, A.; Basheer, C.; Adio, S.; Alhooshani, K.; Lee, H. Removal of haloethers, trihalomethanes and haloketones from water using *Moringa oleifera* seeds. *Int. J. Environ. Sci. Technol.* **2016**, *13*, 2609–2618. [CrossRef]

102. Muthuraman, G.; Sasikala, S. Removal of turbidity from drinking water using natural coagulants. *J. Ind. Eng. Chem.* **2014**, *20*, 1727–1731. [CrossRef]

103. Orwa, C.; Mutua, A.; Kindt, R.; Jamnadass, R.; Anthony, S. *Agroforestree Database: A Tree Species Reference and Selection Guide Version 4.0 [CD-ROM]*; World Agro-forestry Centre: Nairobi, Kenya, 2009; pp. 335–336.

104. Aruna, M.; Srilatha, N. Water clarification using *Moringa oleifera* Lam. seed as a natural coagulant. *Curr. Biot.* **2012**, *5*, 472–486.

105. Egbuikwem, P.; Sangodoyin, A. Coagulation efficacy of *Moringa oleifera* seed extract compared to alum for removal of turbidity and *E. coli* in three different water sources. *Eur. Int. J. Sci. Technol.* **2013**, *2*, 13–20.

106. Preston, K.; Lantagne, D.; Kotlarz, N.; Jellison, K. Turbidity and chlorine demand reduction using alum and moringa flocculation before household chlorination in developing countries. *J. Water Health* **2010**, *8*, 60–70. [CrossRef] [PubMed]

107. Baig, T.H.; Garcia, A.E.; Tiemann, K.J.; Paso, E. Adsorption of heavy metal ions by the biomass of *Solanum elaeagnifolium* (silverleaf night-shade). In Proceedings of the 1999 conference on Hazardous Waste Research, Missouri, MO, USA, 24–27 May 1999.

108. Acheampong, M.A.; Pereira, J.P.C.; Meulepas, R.J.W.; Lens, P.N.L. Kinetics modelling of Cu(II) biosorption on to coconut shell and *Moringa oleifera* seeds from tropical regions. *Environ. Technol.* **2012**, *33*, 409–417. [CrossRef] [PubMed]

109. Sharma, P.; Kumari, P.; Srivastava, M.M.; Srivastava, S. Ternary biosorption studies of Cd(II), Cr(III) and Ni(II) on shelled *Moringa oleifera* seeds. *Bioresour. Technol.* **2007**, *98*, 474–477. [CrossRef] [PubMed]

110. Sharma, P. Removal of Cd(II) and Pb(II) from aqueous environment using *Moringa oleifera* seeds as biosorbent: A low cost and ecofriendly technique for water purification. *Trans. Indian Inst. Met.* **2008**, *61*, 107–110. [CrossRef]

111. Kumari, P.; Sharma, P.; Srivastava, S.; Srivastava, M.M. Biosorption studies on shelled *Moringa oleifera* Lamarck seed powder: Removal and recovery of arsenic from aqueous system. *Int. J. Miner. Process.* **2006**, *78*, 131–139. [CrossRef]

112. Kumar, V.K.; Rubha, M.N.; Manivasagan, M.; Babu, R.; Balaji, P. *Moringa oleifera*—The Nature's Gift. *Univ. J. Environ. Res. Technol.* **2012**, *2*, 203–209.

113. Sajidu, S.M.; Henry, E.M.T.; Kwamdera, G.; Mataka, L. Removal of lead, iron and cadmium ions by means of polyelectrolytes of the *Moringa oleifera* whole seed kernel. *WIT Trans. Ecol. Environ.* **2006**, *80*, 1–8.

114. Abirami, M.; Rohini, C. Comparative study on the treatment of turbid water using *Moringa oleifera* and Alum as coagulants. In Proceedings of the International Conference on Emerging Trends in Engineering, Science and Sustainable Technology, Thudupathi, India, 5–6 April 2018.

115. Maina, I.W.; Obuseng, V.; Nareetsile, F. Use of *Moringa oleifera* (Moringa) seed pods and *Sclerocarya birrea* (Morula) nut shells for removal of heavy metals from wastewater and borehole water. *J. Chem.* **2016**. [CrossRef]

116. Shan, T.C.; Al Matar, M.; Makky, E.A.; Ali, E.N. The use of *Moringa oleifera* seed as a natural coagulant for wastewater treatment and heavy metals removal. *Appl. Water Sci.* **2017**, *7*, 1369–1376. [CrossRef]

117. Pagnanelli, F.; Mainelli, S.; Veglio, F.; Toro, L. Heavy metal removal by olive pomace: Biosorbent characterization and equilibrium modeling. *Chem. Eng. Sci.* **2003**, *58*, 4709–4717. [CrossRef]

118. Reddy, D.H.K.; Ramana, D.K.V.; Seshaiah, K.; Reddy, A.V.R. Biosorption of Ni(II) from aqueous phase by *Moringa oleifera* bark, a low cost biosorbent. *Desalination* **2011**, *268*, 150–157. [CrossRef]

119. Reddy, D.H.K.; Seshaiaha, K.; Reddy, A.V.R.; Leec, S.M. Optimization of Cd(II), Cu(II) and Ni(II) biosorption by chemically modified *Moringa oleifera* leaves powder. *Carbohydr. Polym.* **2012**, *88*, 1077–1086. [CrossRef]

120. Ogle, H. Disease Management: Chemicals. Available online: https://www.appsnet.org/Publications/Brown_Ogle/24%20Control-chemicals%20(HJO).pdf (accessed on 7 August 2018).

121. El-Mougy, N.; Abdel-Kader, L. Vegetables root rot disease management by an integrated control measures under greenhouse and plastic houses conditions in Egypt—A review. *Int. J. Eng. Innov. Technol.* **2012**, *6*, 241–248.

122. Najar, A.G.; Anwar, A.; Masoodi, L.; Khar, M.S. Evaluation of native biocontrol agents against *Fusarium solani* f. sp. melongenae causing wilt disease of brinjal in Kashmir. *J. Phytol.* **2011**, *3*, 31–34.

123. Satish, S.; Mohana, D.C.; Raghavendra, M.P.; Raveesha, K.A. Antifungal activity of some plant extracts against important seed borne pathogens of *Aspergillus* sp. *J. Agric. Technol.* **2007**, *3*, 109–119.

124. Kuri, S.K.; Islam, R.M.; Mondal, U. Antifungal potentiality of some botanical extracts against important seedborne fungal pathogen associated with brinjal seeds, *Solanum melongena* L. *J. Agric. Technol.* **2011**, *7*, 1139–1153.

125. Mariani, C.; Braca, A.; Vitalini, S.; Tommasi, N.D.; Visioli, F.; Fico, G. Flavonoid characterization and in vitro antioxidant activity of *Aconitum anthora* L. (Ranunculaceae). *Phytochemistry* **2008**, *69*, 1220–1226. [CrossRef] [PubMed]

126. Saavedra Gonzalez, Y.R.; van der Maden, E.C.L.J. *Opportunities for Development of the Moringa Sector in Bangladesh: Desk-Based Review of the Moringa Value Chains in Developing Countries and End-Markets in Europe*; Centre for Development Innovation: Wageningen, WD, USA, 2015.

127. Thanaa, S.H.M.; Kassim, N.E.; AbouRayya, M.S.; Abdalla, A.M. Influence of foliar application with moringa (*Moringa oleifera* L.) leaf extract on yield and fruit quality of Hollywood plum cultivar. *J. Hortic.* **2017**, *4*. [CrossRef]

128. Balogun, S.O.; Idowu, A.A.; Ojiako, F.O. Evaluation of the effects of four plant materials and Fernazzan D on the mycelial growth of *Aspergillus flavus* isolated from stored maize grains. *Plant Sci.* **2004**, *4*, 105–114.

129. Makkar, H.P.S.; Becker, K. Nutrients and antiquality factors in different morphological parts of the *Moringa oleifera* tree. *J. Agric. Sci.* **1997**, *128*, 311–322. [CrossRef]

130. Hussain, S.; Malik, F.; Mahmood, S. Review: An exposition of medicinal preponderance of *Moringa oleifera* (Lank.). *Pak. J. Pharm. Sci.* **2014**, *27*, 397–403. [PubMed]

131. Goss, M.; Mafongoya1, P.; Gubba, A. *Moringa oleifera* extracts effect on *Fusarium solani* and *Rhizoctonia solani* growth. *Asian Res. J. Agric.* **2017**, *6*, 1–10. [CrossRef]

132. Chen, M. Elucidation of bactericidal effects incurred by *Moringa oleifera* and Chitosan. *J. U.S. SJWP* **2009**, *4*, 65–79.

133. Viera, G.H.F.; Mourão, J.A.; Ângelo, Â.M.; Costa, R.A.; Vieira, R.H.S.D.F. Antibacterial effect (in vitro) of *Moringa oleifera* and *Annona muricata* against Gram positive and Gram negative bacteria. *Rev. Inst. Med. Trop. Sao Paulo* **2010**, *52*, 129–132. [CrossRef] [PubMed]

134. Jahn, S.A.; Musnad, H.A.; Burgstaller, H. The tree that purifies water: Cultivating multipurpose Moringaceae in the Sudan. *Unasylva* **1986**, *38*, 23–28.

135. Holetz, F.B.; Greisiele, L.P.; Neviton, R.S.; Diógenes, A.G.C.; Celso, V.N.; Benedito, P.D.F. Screening of some plants used in the Brazilian folk medicine for the treatment of infectious diseases. *Mem. Inst. Oswaldo Cruz* **2002**, *97*, 1027–1031. [CrossRef] [PubMed]

136. Bukar, A.; Uba, A.; Oyeyi, T.I. Antimicrobial profile of *Moringa oleifera* lam. Extracts against some food born microorganism. *Bayero J. Pure Appl. Sci.* **2010**, *3*, 43–48.

137. Abalaka, M.E.; Daniyan, S.Y.; Oyeleke, S.B.; Adeyemo, S.O The antibacterial evaluation of *Moringa oleifera* leaf extracts on selected bacterial pathogens. *J. Microbiol. Res.* **2012**, *2*, 1–4. [CrossRef]

138. Food and Agricultural Organization (FAO). Moringa Traditional Crop of the Month. Available online: http://www.fao.org/traditional-crops/moringa/en/2014 (accessed on 7 August 2018).

139. Jabeen, R.; Shahid, M.; Jamil, A.; Ashraf, M. Microscopic of the antimicrobial activity of seed extracts of Moringa. *Pak. J. Bot.* **2008**, *40*, 1349–1358.

140. Chollom, S.C.; Agada, G.O.A.; Gotep, J.G.; Mwankon, S.E.; Dus, P.C.; Bot, Y.S.; Nyango, D.Y.; Singnap, C.L.; Fyaktu, E.J.; Okwori, A.E.J. Investigation of aqueous extract of *Moringa oleifera* lam seed for antiviral activity against newcastle disease virus in ovo. *J. Med. Plants Res.* **2012**, *6*, 3870–3875. [CrossRef]

141. Donkor, A.; Glover, R.; Addae, K.; Kubi, K. Estimating the nutritional value of the leaves of *Moringa oleifera* on poultry. *Food Nutr. Sci.* **2013**, *4*, 1077–1083.

142. Ayssiwede, S.B.; Zanmenou, J.C.; Issa, Y.; Hane, M.B.; Dieng, A.; Chrysostome, C.A.A.M.; Houinato, M.R.; Hornick, J.L.; Missohou, A. Nutrient composition of some unconventional and local feed resources available in Senegal and recoverable in indigenous chickens or animal feeding. *Pak. J. Nutr.* **2011**, *10*, 707–717. [CrossRef]

143. Olugbemi, T.S.; Mutayoba, S.K.; Lekule, F.P. *Moringa oleifera* leaf meal as a hypocholesterolemic agent in laying hen diets. *Livest. Res. Rural Dev.* **2010**, *22*, 84.

144. Abou-Elezz, F.; Sarmiento-Franco, L.; Santos-Ricalde, R.; Solorio-Sanchez, F. Nutritional effects of dietary inclusion of *Leucaena leucocephala* and *Moringa oleifera* leaf meal on Rhode Island Red hens' performance. *Cuban J. Agric. Sci.* **2011**, *45*, 163–169.

145. Kakengi, A.; Kaijage, J.; Sarwatt, S.; Mutayoba, S.; Shem, M.; Fujihara, T. Effect of *Moringa oleifera* leaf meal as a substitute for sunflower seed meal on performance of laying hens in Tanzania. *Int. J. Poult. Sci.* **2007**, *9*, 363–367.

146. Safa, M.; Tazi, E. Effect of feeding different levels of *Moringa oleifera* leaf meal on the performance and carcass quality of broiler chicks. *Int. J. Sci. Res.* **2014**, *3*, 147–151.

147. Gadzirayi, C.T.; Masamha, B.; Mupangwa, J.F.; Washaya, S. Performance of broiler chickens fed on mature *Moringa oleifera* leaf meal as a protein supplement to soyabean meal. *Int. J. Poult. Sci.* **2012**, *11*, 5–10. [CrossRef]

148. Kaijage, J.; Sarwatt, S.; Mutayoba, S. *Moringa oleifera* leaf meal can improve quality characteristics and consumer preference of marketable eggs. *Numer. Proc. Pap.* **2004**, 126–129.

149. Etalem, T.; Getachew, A.; Mengistu, U.; Tadelle, D. *Moringa oleifera* leaf meal as an alternative protein feed ingredient in broiler ration. *Int. J. Poult. Sci.* **2013**, *12*, 289–297.

150. Jung, S.; Choe, J.; Kim, B.; Yun, H.; Kruk, Z.; Jo, C. Effect of dietary mixture of garlic acid and linoleic acid on antioxidative potential and quality of breast meat from broilers. *Meat Sci.* **2010**, *86*, 520–526. [CrossRef] [PubMed]

151. Pennington, J.; Fisher, R. Classification of fruits and vegetables. *J. Food Compos. Anal.* **2009**, *22*, 23–31. [CrossRef]
152. Mutayoba, S.; Mutayoba, B.; Okot, P. The performance of growing pullets fed diets with varying energy and leucaena leaf meal levels. *Livest. Res. Rural Dev.* **2003**, *15*, 350–357.
153. Ebenebe, C.; Anigbogu, C.; Anizoba, M.; Ufele, A. Effect of various levels of moringa leaf meal on the egg quality of Isa Brown Breed of layers. *Adv. Life Sci. Technol.* **2013**, *14*, 1–6.
154. Adeniji, A.; Lawal, M. Effects of replacing groundnut cake with *Moringa oleifera* leaf meal in the diets of grower rabbits. *Int. J. Mol. Vet. Res.* **2012**, *2*, 8–13.
155. Gadzirayi, C.T.; Mupangwa, J.F. The nutritive evaluation and utilisation of *Moringa oleifera* in indigenous and broiler chicken production: A review. *Greener J. Agric. Sci.* **2014**, *14*, 15–21.
156. Lu, W.; Wang, J.; Zhang, H.J.; Wu, S.G.; Qi, G.H. Evaluation of *Moringa oleifera* leaf in laying hens: Effects on laying performance, egg quality, plasma biochemistry and organ histopathological indices. *Ital. J. Anim. Sci.* **2016**, *15*, 658–665. [CrossRef]
157. Ola-Fadunsin, S.D.; Ademola, I.O. Direct effects of *Moringa oleifera* Lam (Moringaceae) acetone leaf extract on broiler chickens naturally infected with *Eimeria species*. *Trop. Anim. Health Prod.* **2013**, *45*, 1423–1428. [CrossRef] [PubMed]
158. Dillard, C.J.; German, J.B. Phytochemicals: Nutraceuticals and human health: A review. *J. Sci. Food Agric.* **2000**, *80*, 1744–1756. [CrossRef]
159. Allen, P.C.; Lydon, J.; Danforth, H.D. Effects of components of *Artemisia annua* on coccidian infections in chickens. *Poult. Sci.* **1997**, *76*, 1156–1163. [CrossRef] [PubMed]
160. Luqman, S.; Srivastava, S.; Kumar, R.; Maurya, A.K.; Chanda, D. Experimental assessment of *Moringa oleifera* leaf and fruit for its antistress, antioxidant and scavenging potential using in vitro and in vivo assays. *Evid.-Based Complement. Altern. Med.* **2012**. [CrossRef] [PubMed]
161. Sonaiya, E.B.; Swan, S.E.J. *Manual Small-Scale Poultry Production Technical Guide*; Food and agriculture organization of the United Nation: Rome, Italy, 2014.
162. Gueye, E.F. Newcastle disease in Family Poultry: Prospects for its control through Ethnovetrinary Medicine. *Livest. Res. Rural Dev.* **2002**, *14*, 25–29.
163. Ferreira, P.M.P.; Farias, D.F.; Oliveira, J.T.A.; Carvalho, A.F.U. *Moringa oleifera*: Bioactive compounds and nutritional potential. *Rev. Nutr.* **2008**, *21*, 431–437. [CrossRef]
164. Davis, L.; Kuttan, G. Suppressive effect of cyclophosphamide- induced toxicity by *Withania somnifera* extract in mice. *J. Ethnopharmacol.* **1998**, *62*, 209–221. [CrossRef]
165. Bouzoubaa, K.; Lemainguer, K.; Bell, J.G. Village chickens as reservoir of *Salmonella pullorum* and *Salmonella gallinarum* in Morocco. *Prev. Vet. Med.* **1992**, *12*, 95–100. [CrossRef]
166. Abiodun, B.S.; Adedeji, A.S.; Taiwo, O.; Gbenga, A. Effects of *Moringa oleifera* root extract on the performance and serum biochemistry of *Escherichia coli* challenged broiler chicks. *J. Agric. Sci.* **2015**, *60*, 505–513. [CrossRef]
167. Patel, J.P. Antibacterial activity of methanolic and acetone extract of some medicinal plants used in India folklore. *Int. J. Phytomed.* **2011**, *3*, 261–269.

![agriculture logo] *agriculture*

MDPI

Article

Effect of Wilting Intensity, Dry Matter Content and Sugar Addition on Nitrogen Fractions in Lucerne Silages

Thomas Hartinger, Nina Gresner and Karl-Heinz Südekum *

Institute of Animal Science, University of Bonn, 53115 Bonn, Germany; thar@itw.uni-bonn.de (T.H.); ngre@itw.uni-bonn.de (N.G.)
* Correspondence: ksue@itw.uni-bonn.de; Tel.: +49-228-73-2287

Received: 24 November 2018; Accepted: 28 December 2018; Published: 5 January 2019

Abstract: Pre-ensiling treatments can significantly influence the composition of lucerne (*Medicago sativa* L.) silages (LS). Besides dry matter (DM) content and availability of water-soluble carbohydrates (WSC), wilting intensity may exert a strong impact on the crude protein (CP; nitrogen [N] × 6.25) fractions. The present study aimed to evaluate the effects of DM level, wilting intensity, and sucrose addition on N compounds and fermentation products in LS. Pure lucerne stand (cultivar Plato) was wilted with either high or low intensity to DM contents of 250 and 350 g kg^{-1}, respectively, and ensiled with or without the addition of sucrose. Non-protein-N (NPN) concentration in LS was affected by all pre-ensiling treatments and with 699 g kg^{-1} CP, NPN was lowest in high-intensity wilted high-DM LS with sucrose addition. No effects were observed on in vitro-estimated concentrations of utilizable CP at the duodenum, a precursor to metabolizable protein. Sucrose addition and higher DM level decreased acetic acid and ammonia-N concentration in the silages. Therefore, the present study demonstrated the beneficial manipulation of CP fractions in LS by high-intensity wilting to higher DM contents and that the provision of WSC may be necessary for sufficient silage fermentation and protein preservation.

Keywords: crude protein; dry matter; lucerne; alfalfa; nitrogen; silage; wilting

---

1. Introduction

Compared with other forage species, lucerne (*Medicago sativa* L.) has a high crude protein (CP; Nitrogen [N] × 6.25) content and depending on its degradability in the rumen, a considerable part of the ruminant's demand for amino acids (AA) can be supplied by feeding lucerne [1]. Preserved as lucerne silage (LS), this forage is continuously available as a component for dairy and beef cattle diets, independently from vegetative growth periods. However, the vast majority of CP in LS is ruminally readily-degradable non-protein-N (NPN), i.e., from 50 up to 87% of total CP [2–4], which can be ascribed to proteolytic activities of lucerne-derived proteases before ensiling, and microbial enzymes during the ensiling process [5]. Legumes are also characterized by low proportions of water-soluble carbohydrates (WSC) [6], which firstly make them difficult to ensile, and secondly also result in silages with minimal concentrations of rapidly fermentable carbohydrates. Solely feeding LS leads to an inefficient microbial N fixation in the rumen [4,7] and consequently high N excretion causing increased environmental pollution. However, substantial N excretion may still occur in mixed LS-based diets because the provision of rapidly fermentable carbohydrates by concentrate is limited due to the risk of rumen acidosis [8]. Therefore, adequately meeting the microbial energy demand for fixing the N arising from the rapid degradation of high NPN amounts in LS is hardly feasible. Consequently, manipulating the CP fractions in LS should be targeted, and in order to improve this fraction, meaning by increasing true protein (TP) concentrations and decreasing low-molecular-weight CP, high-intensity wilting, i.e., with high solar radiation, may be an effective option that to date has not received much attention. Rapid drying should inactivate plant-derived proteases, whose functions rely on water, and consequently stabilize TP content of lucerne plants. Likewise, a previous study by Edmunds et al. [9] already showed that high-intensity wilting results in higher TP percentages in grass silages. Thus, we hypothesized that high-intensity wilting alone or in combination with further treatments may influence the CP composition in LS and decrease proteolysis during ensiling. Because lucerne contains limited amounts of WSC [6], the effect of sucrose addition before ensiling on the N fractions in LS was further tested, particularly as there is clear evidence for decreased ammonia-N concentration in glucose- and fructose-added LS [10] and a more stable silage fermentation in general [11]. Therefore, the objective of the present study was to evaluate N fractions in LS wilted with different intensities to DM contents of 250 or 350 g kg^{-1} and with or without the addition of sucrose. The hypothesis was that the highest TP preservation would occur in those LS, which received high-intensity wilting to 350 g kg^{-1} and with sucrose addition.

2. Materials and Methods

2.1. Preparation of Silages

The procedure for the preparation of the LS was adopted from Edmunds et al. [9] and partly modified as described in the following. On the 19th of July 2016, the third cut of a one hectare pure lucerne stand (cultivar Plato) at the early bud stage of maturity was harvested using a disc mower without a mechanical conditioner at 10 cm stubble height at the Educational and Research Centre Frankenforst of the Faculty of Agriculture, University of Bonn (Königswinter, Germany, 7° 12′ 22′′ E; 50° 42′ 49′′ N). The harvested material was immediately collected from the field and equally spread on either black plastic in the sun (high-intensity wilting; HI) or on white plastic in the shade (low-intensity wilting; LI). The lucerne layers on each plastic had a thickness of approximately 10 cm to ensure sufficient and consistent exposure of the entire plant material to the solar radiation. Immediately, a composite sample was taken and stored at −20 °C for later analysis. This composite sample consisted of 20 single samples that were taken from different places of the lucerne layers on the white and the black plastic, respectively. During silage preparation, the sky was clear, and the weather conditions were sunny with a relative humidity of 59%, a maximum temperature of 32 °C and 15 h of sunshine during the day and a minimum temperature of 20 °C during the night. The plant material was wilted to DM levels (DML) of 250 and 350 g kg^{-1}, respectively, and ensiled either without or with sucrose addition (SU) of 125 g kg^{-1} DM. The amount of added sucrose was chosen as it constitutes the difference between the average WSC content of lucerne with 65 g kg^{-1} DM and perennial ryegrass (*Lolium perenne* L.) with 190 g kg^{-1} DM [12], which is good to ensile [13]. The compaction of the lucerne at ensiling was calculated according to the recommendations of the Federal Working Group for Forage Preservation (Bundesarbeitskreis Futterkonservierung; [12]) in Germany with 190.4 (±2.3) kg DM m^{-3} for low-DM LS and 215.8 (±4.6) kg DM m^{-3} for high-DM LS. The lucerne was ensiled in duplicate in 60 l plastic containers and stored for 120 days. Thus, eight different silage treatments were finally prepared, which are referred to as: 250HISU, 250HI, 250LISU, 250LI, 350HISU, 350HI, 350LISU, and 350LI. The required wilting durations were 2.5 h for 250HISU and 250HI, 4.0 h for 250LISU and 250LI, 7 h for 350HISU and 350HI and 22 h for 350LISU and 350LI.

2.2. Basic Analysis

After 120 days, the two plastic containers of each LS were pooled and three composite samples, each comprising 20 single samples from different spots of the silage heap, were taken and checked for the presence of mould or any other signs of spoilage. All composite samples were thoroughly mixed and 800 g fresh matter of each were freeze-dried and ground successively using 3 mm and then 1 mm sieves (SM 100, Retsch, Haan, Germany). These samples were used for the following analyses, except fermentation pattern analysis, which was conducted with two subsamples (50 g) of each LS that were immediately taken after silo opening and stored at −20 °C.

The proximate analyses were conducted in accordance with the Association of German Agricultural Analytic and Research Institutes (VDLUFA; [14]). The DM content was determined by drying the fresh silages overnight at 60 °C and subsequently at 105 °C for at least 3 h (method 3.1). Using the equation from Weissbach and Kuhla [15], DM was corrected for the loss of volatile compounds that occur during drying. Crude protein was determined by the Kjeldahl method (method 4.1.1) using a Vapodest 50s carousel (Gerhardt, Königswinter, Germany) and multiplying N by 6.25. Proportions of neutral detergent fibre assayed with a heat stable amylase and expressed exclusive of residual ash (aNDFom), acid detergent fibre expressed exclusive of residual ash (ADFom), and acid detergent lignin (ADL) were determined in accordance with methods 6.5.1, 6.5.2, and 6.5.3, respectively.

2.3. Crude Protein Fractionation and Amino Acid Analysis

The CP fractionation was performed according to the Cornell Net Carbohydrate and Protein System [16], following recommendations and standardizations of Licitra et al. [17]. Briefly, five fractions (all expressed as $g\ kg^{-1}$ CP; A, B1, B2, B3, and C) were obtained; fraction A represents NPN, fraction B1 represents rapidly ruminally degradable TP, fraction B2 represents moderately ruminally degradable TP, fraction B3 represents slowly ruminally degradable TP and fraction C represents indigestible TP. Fraction A is the difference between total CP and TP, which precipitates in tungstic acid. Fraction B1 is the difference between total TP and borate-phosphate-buffer-insoluble TP. Fraction B2 is borate-phosphate-buffer-insoluble TP minus neutral-detergent-insoluble TP and fraction B3 is the difference between neutral-detergent-insoluble TP and acid detergent-insoluble TP. Fraction C is acid-detergent-insoluble TP. Subsequently, total TP concentrations ($g\ kg^{-1}$ CP) of samples were calculated by 1000 minus fraction A.

The contents of free AA and total AA (sum of peptide-bound and free AA), including gamma-aminobutyric acid (GABA), were determined by ion-exchange chromatography according to the Commission Regulation (EC) No. 152/2009 of the European Communities [18]. This method is not valid for the determination of tryptophan and cannot differentiate between D and L forms of AA. Briefly, free AA were extracted with diluted hydrochloric acid and co-extracted nitrogenous macromolecules were precipitated with sulfosalicylic acid and removed by filtration before the free AA determination by ninhydrin reaction with spectrophotometric detection at 570 nm. The procedure for total AA determination depended on AA under investigation. Prior to hydrolysis, Cys and Met were oxidized with a performic acid-phenol mixture to cysteic acid and methionine sulphone, respectively, whereas Tyr was determined in unoxidized samples only. All remaining AA were determined in either the oxidized or unoxidized sample. Samples were then hydrolyzed with hydrochloric acid and determined by ninhydrin reaction using spectrophotometric detection at 570 nm or 440 nm for Pro.

2.4. Modified Hohenheim Gas Test

In order to estimate utilizable CP at the duodenum (uCP), the modified Hohenheim gas test [19,20] was conducted as outlined in detail by Edmunds et al. [21]. Briefly, ruminal fluid was collected before morning feeding from two rumen-fistulated sheep receiving a 1:1 grass hay-pelleted compound maintenance ration twice daily. An amount corresponding to 200 mg DM of each sample was incubated in duplicate in each of two runs in 30 mL of ruminal fluid-buffer solution for 8 and 48 h, as recommended for forages [22]. At the end of these incubation periods, syringe contents were analyzed for ammonia-N applying a Vapodest 50s carousel and uCP was calculated using the following equation:

$$uCP\ (g\ kg^{-1}\ DM) = ((ammonia\text{-}N_{blank} + N_{sample} - ammonia\text{-}N_{sample})/sample\ weight\ (mg\ DM)) \times 6.25 \times 1000,$$

where ammonia-N is in $mg\ 30\ mL^{-1}$, blank refers to the ruminal fluid-buffer solution without sample substrate, sample refers to the ruminal fluid-buffer solution with sample substrate, N_{sample} is N added to the syringe through the sample substrate (mg), and sample weight is the amount of sample substrate (mg DM) weighed into the syringe. When using a live product such as ruminal fluid, small biological fluctuations among runs are inevitable. To correct for this a protein standard provided by the University of Hohenheim was analyzed with every run. The standard was a concentrate mixture of (kg^{-1} DM) 450 g rapeseed meal, 300 g faba beans, and 250 g molasses sugar beet pulp, and had a CP content of 254 g kg^{-1} DM. The correction follows the same method as for gas production [23] whereby the mean uCP value for the standard, provided by the University of Hohenheim for 8 or 48 h, is divided by the recorded value of the standard for that run and all other samples are multiplied by the resulting correction factor. Whole runs were repeated if the correction factor, for either incubation time, lay outside the range of 0.9–1.1. The hay and concentrate standards typically used for correcting gas production were also included in the incubation, not only to correct gas production values, but to ensure the ruminal fluid solution followed typical fermentation. After the correction of obtained uCP, values from the incubation times were plotted against a log ((ln) time) scale and the resulting regression equation was used to calculate the effective uCP at passage rates of 0.02, 0.05, and 0.08 hr^{-1}, which are referred to as uCP2, uCP5, and uCP8, respectively. These passage rates represent the ruminal digesta flow, including the solid and liquid phase, in animals with different production levels [24].

2.5. Fermentation Pattern Analysis

Subsamples (50 g) of all silages were used for fermentation pattern analysis. Procedures, as well as detection limits, are described in detail by Brüning et al. [25]. Briefly, a cold-water extract was prepared from all samples by blending the frozen substrate with 200 mL distilled water and 1 mL toluene and refrigerated overnight at 4 °C. Extracts were then filtered using MN 615 filter paper (Macherey-Nagel, Düren, Germany) and subsequently microfiltered (Minisart RC, 0.45 μm pore size; Sartorius, Göttingen, Germany). Ammonia-N concentration was analyzed colorimetrically based on the Berthelot reaction [26]. The pH of the extracts was determined potentiometrically and lactic acid concentration was analyzed by high-performance liquid chromatography with refractive index detection in accordance with Weiß and Kaiser [27]. Volatile fatty acids, alcohols (methanol, ethanol, propanol, butanol, 2,3-butanediol), ethyl lactate, ethyl acetate, propyl acetate, and acetone were determined by gas chromatography with flame ionization detection [28,29]. The concentrations of WSC were determined using the anthrone method [30].

2.6. Statistical Analysis

Statistical analysis was performed with the GLM procedure of SAS version 9.3 (SAS Institute Inc., Cary, NC, USA) using the following model: $Y = \mu + a_i + b_j + c_k + e_{ejk}$ where μ is the mean, a_i is the effect of the SU, b_j is the effect of the wilting intensity (WI), c_k is the effect of the DML and e_{ijk} is the residual error. The significance level was set at $\alpha = 0.05$. In order to test for interactions, field replicates would have been necessary [31], which were not available in the present study. As a consequence, silos were pooled to avoid an artificially created variation and only the main effects were tested. Particularly due to the limited extent of the present study, we preferred to cautiously draw conditional conclusions from a smaller data set as recommended by Lowry [32].

3. Results

3.1. General Chemical Composition

As shown in Table 1, DM content was affected by SU and was slightly higher in SU LS. Concerning the CP content, effects of all three pre-ensiling treatments could be observed, whereby CP proportions ranged from 188 to 219 g kg^{-1} DM and were higher in LS without SU, LI, and 250DML, respectively. The SU treatment also affected the fibre fractions aNDFom and ADFom, which were lower in SU LS. No treatment factor had an effect on ADL.

Table 1. Effect of sucrose addition (SU), wilting intensity (WI), and dry matter (DM) level (DML) on DM content (g kg^{-1}), crude protein content (g kg^{-1} DM), and fibre fractions (g kg^{-1} DM) in lucerne silages (fresh lucerne values are provided as ease for comparison).

Silage	DM	CP	aNDFom	ADFom	ADL
Fresh lucerne	213.1	213	431	340	91
250HISU	254.8	195	458	322	88
250HI	240.5	215	463	364	88
250LISU	255.0	198	422	325	87
250LI	246.8	219	429	355	86
350HISU	344.5	188	416	325	88
350HI	340.0	211	446	338	90
350LISU	346.8	195	390	312	96
350LI	339.0	213	421	336	95
Results of statistical analyses					
SEM	18	4	9	6	1
SU	**	**	*	*	NS
WI	NS	*	NS	NS	NS
DML	**	*	NS	NS	NS

250HISU = 250 g kg^{-1}, high-intensity wilting and sucrose addition; 250HI = 250 g kg^{-1}, high-intensity wilting and no sucrose addition; 250LISU = 250 g kg^{-1}, low-intensity wilting and sucrose addition; 250LI = 250 g kg^{-1}, low-intensity wilting and no sucrose addition; 350HISU = 350 g kg^{-1}, high-intensity wilting and sucrose addition; 350HI = 350 g kg^{-1}, high-intensity wilting and no sucrose addition; 350LISU = 350 g kg^{-1}, low-intensity wilting and sucrose addition; 350LI = 350 g kg^{-1}, low-intensity wilting and no sucrose addition; DM = Dry matter; CP = Crude protein; aNDFom = Neutral detergent fibre after incineration and amylase treatment; ADFom = Acid detergent fibre after incineration; ADL = Acid detergent lignin; SEM = Standard error of the mean (without consideration of fresh lucerne); NS = not significant; * = $p < 0.05$; ** = $p < 0.01$.

3.2. Crude Protein Fractions and Amino Acids

The CP fractionation revealed various differences between the eight silage treatments (Table 2). Non-protein N was the largest CP fraction in all LS but was more than 110 g kg CP^{-1} higher for 250LI when compared to 350HISU. Likewise, NPN (fraction A) was affected by all three factors, i.e., SU, WI, and DML, with increased proportions in 250DML silages. Both HI and SU decreased the NPN proportion in LS. Moderately ruminally degradable TP (fraction B2) was the second largest fraction and highest in silages with 350DML and SU. As with NPN, the largest difference for moderately ruminally degradable TP was found between 250LI and 350HISU. Rapidly (fraction B1) and slowly ruminally degradable TP (fraction B3), as well as indigestible TP (fraction C), were present in small proportions of total CP and slowly ruminally degradable TP was partly not quantifiable. Thus, the effects of SU and DML on these fractions are negligible. Total TP was calculated by subtracting NPN (fraction A) from total CP. Consequently, 250LI had the lowest TP content and, except for 350LI, was clearly separated from 350DML silages.

Table 2. Effect of sucrose addition (SU), wilting intensity (WI), and dry matter level (DML) on crude protein (CP) fractions (g kg^{-1} CP) and true protein content (g kg^{-1} CP) in lucerne silages (fresh lucerne values are provided as ease for comparison).

Silage	Crude Protein Fraction †					
	A	B1	B2	B3	C	TP
Fresh lucerne	259	289	383	27	42	741
250HISU	772	13	174	0	54	228
250HI	799	6	154	0	53	201
250LISU	782	11	16	0	47	218
250LI	812	11	139	0	58	188
350HISU	699	6	251	2	49	301
350HI	744	6	206	0	49	256
350LISU	718	3	253	2	47	282
350LI	779	7	182	0	46	221
Results of statistical analyses						
SEM	14	1	27	0	1	14
SU	**	NS	**	*	NS	**
WI	*	NS	NS	NS	NS	*
DML	**	NS	**	*	#	**

† According to the Cornell Net Carbohydrate and Protein system [16]; 250HISU = 250 g kg^{-1}, high-intensity wilting and sucrose addition; 250HI = 250 g kg^{-1}, high-intensity wilting and no sucrose addition; 250LISU = 250 g kg^{-1}, low-intensity wilting and sucrose addition; 250LI = 250 g kg^{-1}, low-intensity wilting and no sucrose addition; 350HISU = 350 g kg^{-1}, high-intensity wilting and sucrose addition; 350HI = 350 g kg^{-1}, high-intensity wilting and no sucrose addition; 350LISU = 350 g kg^{-1}, low-intensity wilting and sucrose addition; 350LI = 350 g kg^{-1}, low-intensity wilting and no sucrose addition; TP = True protein; SEM = Standard error of the mean (without consideration of fresh lucerne); NS = not significant; # = $0.05 < p < 0.1$; * = $p < 0.05$; ** = $p < 0.01$.

Both SU and DML affected several AA concentrations determined as peptide-bound and free AA, whereas only a few were influenced by WI (Table 3). Concentrations of Thr, Arg, Ser, Asp, and Glu were increased by SU, whereas it decreased Ile, Leu, Val, and Ala. Besides, a strong tendency ($p = 0.06$) for increased Lys concentrations in SU LS were observed. The HI treatment decreased the concentrations of Ile, and Val. The 350DML treatment increased the concentrations of Cys, Lys, Thr, Arg, His, Ser, Pro, Asp, and Glu. In contrast, concentrations of Ile, Leu, Val, and Ala were decreased in high-DM LS.

The DML treatment affected free AA more than SU or WI (Table 4) and HI tended to decrease Ile concentrations ($p = 0.09$). The SU treatment increased the concentration of free Thr and tended to increase free Glu ($p = 0.07$), whereas it decreased free Ile, Leu, Val, and Ala. The 350DML LS showed higher concentrations of free Lys, Thr, Pro, Asp, Glu as well as free His that was not detectable in 250DML LS. Free Ile, Leu, Val, and Ala were reduced in 350DML LS. Regarding the amount of total free AA, SU decreased total free AA, whereas no influence of other pre-ensiling treatments was observed. Moreover, 350DML and SU reduced the concentrations of free and total GABA (Tables 3 and 4).

Table 3. Effect of sucrose addition (SU), wilting intensity (WI), and dry matter (DM) level (DML) on contents (g kg^{-1} DM) of total amino acids (AA; the sum of peptide-bound and free AA) and gamma-aminobutyric acid (GABA) in lucerne silages.

AA	Ala	Arg	Asp	Cys	Glu	Gly	His	Ile	Leu	Lys	Met	Phe	Pro	Ser	Thr	Val	GABA
250HISU	21.1	2.3	11.3	1.2	10.2	9.6	1.9	9.5	14.8	3.2	3.2	9.3	6.2	3.2	3.6	11.8	10.3
250HI	29.4	1.7	4.2	0.7	5	3.5	1.3	10.3	16.5	2.5	2.2	8.5	2.1	2.2	2	13.3	16.7
250LISU	22.3	2.1	10.6	1.3	10.3	9.7	2	9.7	15	3.1	3.2	9.5	6	2.9	3	12	10.7
250LI	28.8	1.7	4.4	0.9	5.5	6	1.6	10.6	16.3	2.6	2.7	8.4	2.3	2.2	2	13.4	16.5
350HISU	14.7	3.1	17.6	0.13	12.5	8.9	3.3	8.9	14.4	7.6	3	9.1	8.7	4.7	6.4	11	8
350HI	19.1	2.3	13	1.4	9.7	10	3.6	10	16	7.2	3.2	9.4	8.4	3	4	12.6	12.5
350LISU	14.4	3.2	19.9	1.4	12.5	9.2	3.5	9.1	14.5	8.6	3.1	9.4	10.3	5.1	6.9	11.7	7.4
350LI	18.1	2.3	14.4	1.4	10.9	10.3	4.4	10.3	16.3	7.8	3.4	10.2	10.9	3.4	4.4	13.2	10.9
SEM	2.02	0.2	1.99	0.16	1.01	0.84	0.4	0.21	0.31	0.95	0.13	0.2	1.18	0.38	0.65	0.31	1.23
						Results of statistical analyses											
SU	**	**	**	NS	**	NS	NS	**	#	NS	NS	NS	NS	**	**	**	**
WI	NS	NS	NS	NS	NS	NS	NS	*	NS	NS	NS	NS	NS	NS	NS	*	NS
DML	**	**	**	#	**	NS	**	**	*	**	NS	NS	**	**	**	*	**

250HISU = 250 g kg^{-1}, high-intensity wilting and sucrose addition; 250HI = 250 g kg^{-1}, high-intensity wilting and no sucrose addition; 250LISU = 250 g kg^{-1}, low-intensity wilting and sucrose addition; 250LI = 250 g kg^{-1}, low-intensity wilting and no sucrose addition; 350HISU = 350 g kg^{-1}, high-intensity wilting and sucrose addition; 350HI = 350 g kg^{-1}, high-intensity wilting and no sucrose addition; 350LISU = 350 g kg^{-1}, low-intensity wilting and sucrose addition; 350LI = 350 g kg^{-1}, low-intensity wilting and no sucrose addition; SEM = Standard error of the mean; NS = not significant; # = $0.05 < p < 0.1$; * = $p < 0.05$; ** = $p < 0.01$.

Table 4. Effect of sucrose addition (SU), wilting intensity (WI), and dry matter (DM) level (DML) on contents (g kg^{-1} DM) of free amino acids (AA) and gamma-aminobutyric acid (GABA) in lucerne silages.

AA	Ala	Arg	Asp	GABA	Glu	Gly	His	Ile	Leu	Lys	Met	Phe	Pro	Thr	Val
250HISU	21.9	0	0.6	9.7	3	6	0	6	10.4	0	2.1	5.9	2.9	1.4	7.8
250HI	32.5	0	0	12.5	0	0.8	0	8	12.8	0	1.2	5.3	0	0.2	10.6
250LISU	23.5	0	0.4	10.1	3.1	6.2	0	6.3	10.7	0	2.2	6.1	2.8	0.9	8.3
250LI	32	0	0.3	12.4	0.5	3.1	0	8.3	12.7	0	1.7	5.3	0.3	0.3	10.6
350HISU	13.4	0	5.4	7.6	3.9	4.7	1	4.8	8.8	3.6	1.5	4.9	4.6	3.7	6.1
350HI	19.8	0	6	9.3	3	6.6	1.7	7.1	11.4	4.1	1.7	5.8	6	2.1	9.1
350LISU	12.7	0	7.4	7.1	4.1	4.6	1.5	4.9	8.7	4.4	1.5	5.1	6.1	4	6.4
350LI	18.4	0	7.5	8.2	4.3	6.5	2.2	7.3	11.5	4.5	1.7	6.5	8.3	2.4	9.4
SEM	2.64	0	1.21	0.72	0.57	0.71	0.32	0.47	0.55	0.79	0.11	0.19	1.03	0.51	0.61
					Results of statistical analyses										
SU	**	NS	NS	**	#	NS	NS	**	**	NS	NS	NS	NS	**	**
WI	NS	NS	NS	NS	NS	NS	NS	#	NS	NS	NS	NS	NS	NS	NS
DML	**	NS	**	**	*	NS	**	**	**	**	NS	NS	*	**	**

250HISU = 250 g kg^{-1}, high-intensity wilting and sucrose addition; 250HI = 250 g kg^{-1}, high-intensity wilting and no sucrose addition; 250LISU = 250 g kg^{-1}, low-intensity wilting and sucrose addition; 250LI = 250 g kg^{-1}, low-intensity wilting and no sucrose addition; 350HISU = 350 g kg^{-1}, high-intensity wilting and sucrose addition; 350HI = 350 g kg^{-1}, high-intensity wilting and no sucrose addition; 350LISU = 350 g kg^{-1}, low-intensity wilting and sucrose addition; 350LI = 350 g kg^{-1}, low-intensity wilting and no sucrose addition; SEM = Standard error of the mean; NS = not significant; # = $0.05 < p < 0.1$; * = $p < 0.05$; ** = $p < 0.01$.

3.3. Modified Hohenheim Gas Test

Irrespective of calculated passage rate, pre-ensiling treatments had no effect on effective uCP values of LS (Table 5). Only uCP8 values tended to be higher for 250DML LS ($p = 0.08$).

Table 5. Effect of sucrose addition (SU), wilting intensity (WI), and dry matter (DM) level (DML) on effective utilizable crude protein at the duodenum (g kg^{-1} DM).

Silage	uCP2	uCP5	uCP8
250HISU	72	109	127
250HI	82	114	131
250LISU	74	112	131
250LI	76	110	128
350HISU	74	105	121
350HI	80	107	121
350LISU	75	108	124
350LI	74	103	118
SEM	1.2	1.3	1.7
Results of statistical analyses			
SU	NS	NS	NS
WI	NS	NS	NS
DML	NS	NS	#

250HISU = 250 g kg^{-1}, high-intensity wilting and sucrose addition; 250HI = 250 g kg^{-1}, high-intensity wilting and no sucrose addition; 250LISU = 250 g kg^{-1}, low-intensity wilting and sucrose addition; 250LI = 250 g kg^{-1}, low-intensity wilting and no sucrose addition; 350HISU = 350 g kg^{-1}, high-intensity wilting and sucrose addition; 350HI = 350 g, high-intensity wilting and no sucrose addition; 350LISU = 350 g kg^{-1}, low-intensity wilting and sucrose addition; 350LI = 350 g kg^{-1}, low-intensity wilting and no sucrose addition; uCP2 = effective utilizable crude protein at the duodenum to passage rate of 0.02 hr^{-1}; uCP5 = effective utilizable crude protein at the duodenum to passage rate of 0.05 hr^{-1}; uCP8 = effective utilizable crude protein at the duodenum to passage rate of 0.08 hr^{-1}; SEM = Standard error of the mean; NS = not significant; # = $0.05 < p < 0.1$.

3.4. Fermentation Pattern

Acetone, 2,3-butandiol, i-valeric acid, n-valeric acid and propyl acetate were not detected in any sample during fermentation pattern analysis. The SU treatment decreased silage pH and ammonia-N concentration, but increased lactic acid concentration as well as ethyl acetate and ethyl lactate (Table 6). Besides, LS without SU tended to have higher concentrations of acetic acid ($p = 0.09$), WSC (P = 0.09), and ethanol ($p = 0.06$). In contrast, WI had no effect on response variables. The 350DML reduced acetic acid as well as methanol concentration and tended to decrease ammonia-N ($p = 0.06$) and propanol ($p = 0.09$) in LS compared to 250 DML (Table 6).

Table 6. Effect of sucrose addition (SU), wilting intensity (WI), and DM level (DML) on lactic acid, volatile fatty acids, ester compounds, alcohols, water-soluble carbohydrates (g kg⁻¹ DM), and ammonia-nitrogen (N; g kg⁻¹ N) in lucerne silages.

Silage	pH	Lactic Acid	Acetic Acid	Propionic Acid	Butyric Acid	Caproic Acid	Ethyl Acetate	Ethyl Lactate	Methanol	Ethanol	Butanol	Propanol	WSC	Ammonia-N
250HISU	4.58	50.6	38.2	0.8	0.7	0.0	0.2	0.1	2.1	7.9	0.1	1.9	10.1	175
250HI	6.12	5.4	42.8	2.1	21.9	0.8	0.1	0.0	2.5	6.2	0.1	0.3	2.3	276
250LISU	4.61	52.4	38.2	1.5	1.5	0.0	0.2	0.1	2.6	8.7	0.1	1.7	10.6	157
250LI	5.85	15.3	48.4	2.0	7.2	0.0	0.1	0.0	3.0	5.4	0.1	0.3	3.1	221
350HISU	4.77	39.7	31.1	1.0	0.6	0.0	0.2	0.1	1.7	6.5	0.0	0.2	17.6	145
350HI	5.81	21.6	34.0	0.3	0.5	0.0	0.1	0.0	2.2	6.3	0.1	0.2	5.0	217
350LISU	4.65	36.2	31.2	0.8	0.3	0.0	0.2	0.1	1.2	5.8	0.0	0.1	46.0	149
350LI	5.73	38.4	31.4	1.3	0.3	0.0	0.1	0.0	1.8	4.3	0.1	0.2	4.8	191
Results of statistical analyses														
SEM	0.24	5.94	2.22	0.2	2.7	0.13	0.02	0.02	0.20	0.49	0.02	0.26	5.12	15.9
SU	**	*	#	NS	NS	NS	**	**	NS	#	NS	NS	#	**
WI	NS	NS	NS	NS	NS	NS	NS	NS	NS	NS	NS	NS	NS	NS
DML	NS	NS	**	NS	NS	NS	NS	NS	*	NS	NS	#	NS	#

WSC = Water-soluble carbohydrates; 250HISU = 250 g kg⁻¹, high-intensity wilting and sucrose addition; 250HI = 250 g kg⁻¹, high-intensity wilting and no sucrose addition; 250LISU = 250 g kg⁻¹, low-intensity wilting and sucrose addition; 250LI = 250 g kg⁻¹, low-intensity wilting and no sucrose addition; 350HISU = 350 g kg⁻¹, high-intensity wilting and sucrose addition; 350HI = 350 g kg⁻¹, high-intensity wilting and no sucrose addition; 350LISU = 350 g kg⁻¹, low-intensity wilting and sucrose addition; 350LI = 350 g kg⁻¹, low-intensity wilting and no sucrose addition; SEM = Standard error of the mean; NS = not significant; # = 0.05 < p < 0.1; * = p < 0.05; ** = p < 0.01.

4. Discussion

4.1. General Chemical Composition

Crude protein contents were lower in SU LS, which likely reflects a dilution caused by the SU. The same may apply to aNDFom and ADFom concentrations. Besides a dilution effect, a stronger acidic hydrolysis of hemicelluloses by acids [33] originating from microbial sucrose metabolism may have occurred during ensiling, consequently causing the lower aNDFom concentrations in SU LS. Nonetheless, proportions of aNDFom and ADFom were in a similar range of other LS [34,35]. Likewise, proportions of fibre fractions in fresh lucerne material were similar to previous findings [3,34,35].

The DML also affected CP content of LS, which was higher in 250 DML LS. This slight difference could have been caused by mechanical losses during harvest and consequently a lower leaf proportion in the ensiled plant material. Therefore, LS with higher DM contents seem to be favorable with regards to CP composition, but should not exceed a certain level to preserve the leaf fraction that dries faster than the stem part and thus is more prone to field losses [36]. The WI treatment also effected CP content, but the effect seems negligible as mean CP difference between LS with HI and LI was only 4 g kg^{-1} DM.

4.2. Crude Protein Fractions

Crude protein fractions in fresh lucerne material were in a typical range for this forage legume, although the present proportion of NPN compounds of 259 g kg^{-1} CP was substantially higher than literature data, i.e., 150 g kg^{-1} CP [3], 170-183 g kg^{-1} CP [37] and 180–190 g kg^{-1} CP [35]. It may be noted that the highest discrepancy in NPN proportion, found between the results of the present study and of Guo et al. [3], might partly also derive from different methodologies. In the present study, tungstic acid was used to precipitate TP, which cuts off peptides of an approximate chain length of more than three AA [17]. Guo et al. [3], however, used trichloroacetic acid to precipitate TP, which cuts off at about 10 AA [17].

The pattern of CP fractions in LS with NPN being the largest, and moderate ruminally degradable TP being the second largest proportion of total CP corresponded to the literature [3,35]. It is notable that NPN contents of present LS were similar to those from Broderick [2] and Seale et al. [10], but higher than values reported by others, for instance, 684 g kg CP^{-1} [3] or 599 g kg CP^{-1} [38] in untreated LS. However, these NPN values were determined after only 35 or 30 days of ensiling, respectively, probably underestimating NPN in LS as intrinsic protease and carboxypeptidase were recently shown to remain largely active for more than 30 days after ensiling [38]. Therefore, NPN values of the present LS, which were stored for 120 days, might provide a more realistic insight and should be considered when comparing different results or designing experiments for silage additive evaluation in LS. In this context, it may be noted that the Federal Working Group for Forage Preservation (Bundesarbeitskreis Futterkonservierung; [12]) recommends at least 90 days of ensiling for any silage-related experiment, e.g., when evaluating the effect of silage additives. Besides, the influence of the cut number should also be taken into account as NPN was 10% higher in third-cut LS when compared to NPN proportion of first-cut LS from the same sward [35]. Likewise, present LS was produced from a third cut, which thus may have been a contributing factor and should be investigated in future studies.

The SU reduced NPN along with increasing moderately ruminally degradable TP, which was likely caused by faster and stronger acidification, consequently suppressing proteolytic microorganisms in the silos [11]. These observations were in accordance with Seale et al. [10] who analyzed the effect of glucose and fructose addition with or without microbial inoculants on LS. However, in Italian ryegrass (*Lolium multiflorum* LAM.), Heron et al. [39] found that plant-derived proteases remained active over a wide pH range, which is also true for lucerne with major endopeptidases having optimum activities at pH 4 [40]. Thus, despite SU treatment and probably rapid acidification, the relevant plant-derived proteolytic activity may have taken place, particularly in 250DML LS.

An effect of WI was found for NPN concentration, which was higher in LI LS. Likewise, high-intensity wilted grass silages had approximately 100 g kg^{-1} CP lower NPN proportions compared to low intensively wilted grass silages [9] and, together with the present observations, demonstrate the TP stabilizing effect of HI treatments. The rapid inhibition of plant-derived proteases, which depend on sufficient water availability [5], may be causative. In this context, Owens et al. [41] produced LS with a DM of 350 g kg^{-1} and observed reduced NPN amounts of approximately 50 g kg^{-1} total N in those LS that needed shorter wilting periods to reach the desired DM, which thus can be ascribed to a higher WI. Likewise, when wilting times were different due to varying levels of shade during wilting, they also observed an increase of NPN with shade, thus substantiating the TP stabilizing effect of an HI treatment. However, it must be considered that although there is evidence for a reduction of plant-derived protease activity by HI treatment, it is very arguable whether plant enzymes were completely deactivated as the moisture loss was only until a DM content of maximal 350 g kg^{-1}. Thus, plant-derived proteases may still have contributed to overall proteolytic processes resulting in the still substantial conversion of TP to NPN in high-intensity wilted LS.

The DML treatment also effected NPN proportion, which was lower in 350DML LS. This confirmed previous findings [9,42] and may be explained by a lower water activity in the silos, consequently reducing microbial metabolism [34]. However, this mechanism should be even more pronounced at DM contents above 500 g kg^{-1} [9]. The effect of DML on slowly ruminally degradable TP may be of marginal importance as this fraction could not be determined in six of eight LS. The higher contents of moderately ruminally degradable TP in 350DML LS, however, may be beneficial regarding the quality of CP that is provided to the animal, meaning a decelerated ruminal CP degradation and therefore potentially improved N utilization by rumen microorganisms. As obtained for the TP proportion of 350HISU, the combination of HI, elevated DML and SU should have limited both plant-derived and microbial CP degradation and thus most effectively stabilized the TP content in the present study.

4.3. Amino Acids

A variety of factors influence the AA composition in silages, including wilting rate, acidification, and the microbial activity in the silo, but also plant-associated factors like tannin concentration or activity of plant proteases [5]. To the authors' knowledge, information about the effects of WI, DML, or SU on AA composition of LS is rare [3,43]. However, as the vast majority of AA is degraded in the rumen, knowledge on AA profiles seems to be more important for feedstuffs with high ruminally undegradable CP [9], which does not apply to the present LS. Though, it is worthy of remark that pre-ensiling treatments clearly effected the AA composition of LS. For instance, the higher proportions of free His, Asp, Lys, Thr, Glu, and Pro in 350DML LS should be the result of reduced microbial activity [5]. Likewise, a similar pattern was observed for total AA. The reducing effect of SU on total free AA content supported the TP preserving effect that was also observed for the distribution of CP fractions and should be caused by rapid acidification [5].

Biogenic amines are predominantly formed during proteolysis in silages [44] and Ohshima and McDonald [45] described the decarboxylation of Glu to GABA during lucerne ensiling, which is reflected by the lower Glu concentrations in the present LS without SU. As summarized by Scherer et al. [44], biogenic amines are associated with lower feed intake and potential impairments to animal health. Thereby, GABA is an important biogenic amine and known to act as a neurotransmitter. It is also involved in the sensation of pain and anxiety as well as neurological diseases [46]. Although there is no clear trend for the effect of GABA on feed intake [44], a negative correlation between feed intake and total amine concentration has been observed [47] and the reduced GABA concentration in LS with 350DML or SU may, therefore, be interpreted as beneficial; particularly also because of potential health risks when biogenic amines would be absorbed by host animals, who, however, are more susceptible under acidotic conditions [48].

4.4. Modified Hohenheim Gas Test

None of the pre-ensiling treatments had an effect on effective uCP values at any calculated passage rate. As Edmunds et al. [9] observed higher uCP values for fast wilted grass silages, the absence of any effect was not expected. Although artificially dried lucerne (90 °C for 3 min) showed reduced effective N degradability and degradation rates in the rumen [49], the WI achieved in the present study may not have been high enough to cause a similar impact. Moreover, the generally limited availability of WSC in all present LS may have prevented an effect on uCP as CP was degraded to ammonia-N, but not reused for microbial protein synthesis in the syringes.

Edmunds et al. [9] found that varying CP contents in grass silages from the same sward can confound the detection of possible effects on uCP as uCP values are calculated from the difference between N content in the syringe, which is determined by the CP content of the sample, and ammonia-N in the syringe. Therefore, these authors recalculated the effective uCP values with an average CP concentration. Thus, as the CP concentrations for the present LS also showed a variance, the effective uCP values were recalculated using the average CP content of LS with and without sucrose, respectively. Thereby, no effects of pre-ensiling treatments on uCP were obtained (data not shown). Moreover, a greater standard deviation might further impair the determination of clear effects. However, the inclusion of additional runs did not reduce standard deviation in the present study and thus were not included in the calculation of effective uCP.

4.5. Fermentation Pattern

The pH values were lower for SU LS, which was reflected by higher lactic acid concentration in these silages. Without the SU treatment, the high buffering capacity of lucerne [7] may have hindered rapid and strong acidification and consequently resulted in higher pH. In this context, the higher ammonia-N contents should also be considered, which can limit the pH drop in silages, as well [50]. Besides, low lactic acid concentration may be caused by metabolic activities of lactate-utilizing lactobacilli [51]. Owens et al. [7] stated a pH below 5.0 as a threshold to maintain forage quality and limit protein degradation in the silo, which thus was only met by SU LS. Likewise, the pH of these silages was within the common range for legume silages at this DM content [50]. Water-soluble carbohydrates [52] as well as total non-structural carbohydrates [7] decrease during wilting of lucerne due to plant enzyme activity and respiration, which are both reduced by moisture loss [53]. Therefore, there should have been less WSC degradation in the plant material undergoing HI treatment and consequently, a stronger pH drop along with increased lactic acid concentrations in HI LS was initially expected. However, the lack of a wilting effect may be explained by the overall very low WSC concentration of lucerne [6], which further was cut in the morning when WSC concentrations are again lower compared to the afternoon [7].

An impact of DM content on silage pH was often described in the literature [34,50]. Thereby, silages with DM contents below 300 g kg^{-1} are extremely susceptible to clostridial fermentation [50], which results in elevated pH values as well as high butyric acid concentrations. In the present study, however, DM content had neither an effect on silage pH nor on butyric acid concentration. Possibly the SU treatment superimposed a potential effect of DML, which is indicated by closer examination of butyric acid concentrations, which were numerically but not statistically significantly higher in 250DML LS. Moreover, the influence of DM content on clostridial fermentation and thus silage pH is more pronounced at DM contents of 400 g kg^{-1} or more [11], which is confirmed by the findings of Santos and Kung [34].

The lower ammonia-N concentration in SU LS further strengthens the assumption that addition of rapidly fermentable carbohydrates better inhibited degradation of nitrogenous compounds in these LS compared to LS without SU, and similar trends have been observed previously [10]. Regarding the impact of DML, ammonia-N concentration is generally higher in wet silages, which corresponds to present findings and is often ascribed to clostridial fermentation [50]. Likewise, reduced ammonia-N contents in LS with high DM contents were also observed by Santos and Kung [34]. Thus, a greater

WI seemed to preserve CP from degradation in the silo. However, according to Wyss et al. [35], ammonia-N proportions lower than 100 g kg^{-1} N is preferable for LS. This threshold was not met in the present study, even for 35HISU. In comparison to fresh-cut lucerne [35], plant material of the present study already showed a higher NPN proportion before ensiling, and may explain the high ammonia-N concentration in LS, irrespectively of applied pre-ensiling treatment. However, the increase of NPN from fresh-cut material to silage material were on the same level in the study by Wyss et al. [35] and the present study. Regarding the high ammonia-N concentration in the present LS, the variation of CP composition between different lucerne cultivars [54] should be considered, as well. Moreover, it can be speculated whether chopping of plant material subsequently supporting silo compression would have increased TP proportions as it was described earlier [50,55]. However, as LS was prepared according to recommended guidelines [12], the latter point may be of minor importance. Besides, higher ammonia-N concentration is assumed to be associated with undesirable metabolites like biogenic amines [50], which is in line with the present observations for higher GABA concentration in 250DML LS.

High concentration of acetic acid is associated with high DM and energy losses [50] as well as considerably reduced ad libitum feed intakes [56]. Compared to the literature [34,35], the acetic acid contents of LS in the present study can be classified as slightly high for 250DML LS and thus would negatively impact their nutritive value. Increased activity of Enterobacteriaceae [57], as well as increased deamination [5], could be causative for acetic acid formation, which is further favoured by high moisture contents [50] and in accordance with the higher acetic acid concentration in 250DML LS. However, the presence of acetic acid is not a disadvantage per se. The average 32 g kg^{-1} DM acetic acid in the 350DML LS, however, might be regarded beneficial as such concentrations have a positive effect on aerobic stability of silages [58].

A butyric acid concentration higher than 5 g kg^{-1} DM indicate elevated clostridial activity and due to high losses of energy, this means diminished energy supply to the animal and, consequently, performance may suffer [50]. This threshold was not exceeded for 350DML LS and only applies to 250HI and 250LI. Likewise, these two treatments also had the highest ammonia-N concentration, which further points to clostridial fermentation [50]. Together with the observation that 250HI and 250LI did also not meet the pH threshold for maintaining forage quality in the silo [7], these two LS should, therefore, be classified as poor-quality silages and potentially spoiled material.

Concerning ethanol, SU tended to increase this alcohol in the silages. Though, ethanol concentration was low for all LS and thus does not indicate elevated yeast metabolism [50,59]. Minor amounts of ethanol can also originate from heterofermentative lactic acid fermentation [60], which can never be fully prevented during ensiling. Weiß and Kalzendorf [52] observed higher concentrations of ethanol and ester compounds in LS with low DM contents and further postulated a positive correlation between ethanol and ester concentration in silages, which is both confirmed by the present findings for ethanol and ethyl lactate as well as ethyl acetate. The effect of esters in silages is not fully clear [47,61], but negative correlations to short-term DM intake were observed earlier [61]. Thus, despite lower NPN proportions in SU LS, the effect of SU on ester occurrence could be regarded as critical.

4.6. General Considerations

Up to now, WI has not received much attention in silage preparation and studies investigating the effect of different WI on silage characteristics are rare. It has been reported that wilting per se effects CP composition, for instance by reducing ammonia-N contents in LS [54]. Thus, applying HI treatments may even be more effective for stabilizing TP content in LS, which is underlined by the present findings that confirm our hypothesis of a TP preserving effect by the HI treatment. A variety of silage additives exists that limit proteolysis in LS [38]. However, they cause costs for acquisition, and in case of organic acids, also for maintenance of corroded machinery and concrete [62]. In contrast, HI treatment does not require additional application systems or further technical equipment and in this regard is an easy to apply tool for improving the quality of on-farm produced protein, and consequently might help to reduce costs for ruminant diets and increase sustainability. Additionally, feed intake of wilted grass silage was increased when compared to non-wilted [63]. In case this also applies to LS, a superior energy and nutrient provision to the animal may be achieved by intensively wilting lucerne to higher DML; provided that mechanical losses during harvest do not exceed the benefits of HI. Regarding the practicality of HI treatment, if possible high solar radiation along with high wind speed should be present during lucerne harvesting. However, this cannot always be guaranteed, which restricts the practicability of HI treatments. To support the effect of rapid dehydration, maceration can be an effective addition to further increase the wilting rate [64] or to compensate weather conditions that may not be as ideal for HI as described in the present study. However, the risk of mechanical losses and thereby associated nutrient changes can be higher when using maceration [64], which needs to be taken into account. Besides, artificial drying treatments are surely a more weather independent option to obtain similar TP preservation [65] as here found for HI in the sun. However, increased production costs due to high energy demands may outweigh the beneficial effects [36] of artificial drying on CP composition.

Concerning the effects of SU on fermentation and CP quality, provision of rapidly fermentable carbohydrates is recommended. Particularly because an inoculation with lactic acid producing bacteria alone may not improve the situation as long as there is not enough easily accessible substrate for lactic acid fermentation [10]. Thus, relating to large scale on-farm conditions, mixing lucerne crop with molasses, crushed cereals, or high WSC forage species before ensiling may constitute a method for equivalently substituting SU treatment in LS. A delayed cutting of lucerne in order to increase non-structural carbohydrates, particularly starch, may not be appropriate as Owens et al. [7] did not find a protein preserving effect in LS differing in WSC content due to different cutting times during the harvest. However, present results revealed concerns about promoted ester and ethanol formation in SU LS that should be kept in mind and require further investigation.

Finally, the pre-ensiling treatment combination of all three factors, i.e., HI treatment to high DML with SU, has the strongest potential to reduce the extent of CP degradation during ensiling, thus improving the protein value and potentially increasing ruminal N retention, particularly when combined with an appropriate carbohydrate source.

5. Conclusions

The effect of WI in silage preparation has not received much attention thus far. However, the present findings underline the importance of HI to limit CP degradation in LS. Therefore, if possible, at harvest, HI should be considered during silage production with lucerne. Regarding the observed effects of SU, providing an additional carbohydrate source to lucerne crop before ensiling is effective to minimize TP degradation and improves silage fermentation quality. However, caution should be paid to volatile organic compounds when operating with SU. Combining an HI treatment to DML of 350 g kg^{-1} with the provision of rapidly fermentable carbohydrates will maintain higher TP proportions along with improving fermentation quality in LS. Otherwise, there is a high chance for poor quality LS that in consequence cannot be fed without the risk of impairing animal performance and health. In order to underpin the present findings and to expand the sparse knowledge on WI, it is necessary to investigate such pre-ensiling treatments over several growth cycles and to further examine if the beneficial effects observed at silage stage can be transferred to rumen fermentation and animals

Author Contributions: Conceptualization, N.G. and K.-H.S.; formal analysis, T.H.; investigation, T.H.; validation, T.H., N.G. and K.-H.S.; writing—original draft preparation, T.H.; writing—review and editing, T.H., N.G. and K.-H.S.; visualization, T.H.; supervision, K.-H.S.; project administration, K.-H.S.; funding acquisition, K.-H.S.

Funding: This research was funded by the "Deutsche Forschungsgemeinschaft" (DFG, German Research Foundation), SU124/33–1.

Acknowledgments: The authors thank K. Weiß and the team at the Corporate Analytical Laboratory, Humboldt-University, Berlin, Germany, for fermentation pattern analysis. The authors thank W. Heimbeck and the team at the laboratory of Evonik Nutrition & Care GmbH, Essen, Germany, for amino acid analysis.

Conflicts of Interest: The authors declare no conflict of interest. The funders had no role in the design of the study; in the collection, analyses, or interpretation of data; in the writing of the manuscript, or in the decision to publish the results.

References

1. Broderick, G.A. Desirable characteristics of forage legumes for improving protein utilization in ruminants. *J. Anim. Sci.* **1995**, *73*, 2760–2773. [CrossRef] [PubMed]
2. Broderick, G.A. Performance of lactating dairy cows fed either alfalfa silage or alfalfa hay as the sole forage. *J. Dairy Sci.* **1995**, *78*, 320–329. [CrossRef]
3. Guo, X.S.; Ding, W.R.; Han, J.G.; Zhou, H. Characterization of protein fractions and amino acids in ensiled alfalfa treated with different chemical additives. *Anim. Feed Sci. Technol.* **2008**, *142*, 89–98. [CrossRef]
4. Coblentz, W.K.; Grabber, J.H. In situ protein degradation of alfalfa and birdsfoot trefoil hays and silages as influenced by condensed tannin concentration. *J. Dairy Sci.* **2013**, *96*, 3120–3137. [CrossRef] [PubMed]
5. Hoedtke, S.; Gabel, M.; Zeyner, A. Protein degradation in feedstuffs during ensilage and changes in the composition of the crude protein fraction (Der Proteinabbau im Futter während der Silierung und Veränderungen in der Zusammensetzung der Rohproteinfraktion). *Übers. Tierernährg.* **2010**, *38*, 157–179. (In German)
6. Lüscher, A.; Mueller-Harvey, I.; Soussana, J.F.; Rees, R.M.; Peyraud, J.L. Potential of legume-based grassland-livestock systems in Europe: A review. *Grass Forage Sci.* **2014**, *69*, 206–228. [CrossRef]
7. Owens, V.N.; Albrecht, K.A.; Muck, R.E.; Duke, S.H. Protein degradation and fermentation characteristics of red clover and alfalfa silage harvested with varying levels of total nonstructural carbohydrates. *Crop Sci.* **1999**, *39*, 1873–1880. [CrossRef]
8. Owens, F.N.; Secrist, D.S.; Hill, W.J.; Gill, D.R. Acidosis in cattle: A review. *J. Anim. Sci.* **1998**, *76*, 275–286. [CrossRef]
9. Edmunds, B.; Spiekers, H.; Südekum, K.-H.; Nussbaum, H.; Schwarz, F.J.; Bennett, R. Effect of extent and rate of wilting on nitrogen components of grass silage. *Grass Forage Sci.* **2014**, *69*, 140–152. [CrossRef]
10. Seale, D.R.; Henderson, A.R.; Pettersson, K.O.; Lowe, J.F. The effect of addition of sugar and inoculation with two commercial inoculants on the fermentation of lucerne silage in laboratory silos. *Grass Forage Sci.* **1986**, *41*, 61–70. [CrossRef]
11. Muck, R.E. Factors influencing silage quality and their implications for management. *J. Dairy Sci.* **1988**, *71*, 2992–3002. [CrossRef]
12. Bundesarbeitskreis Futterkonservierung. *Praxishandbuch Futter- und Substratkonservierung*; vollst. überarb. Aufl. 2011; DLG-Verlag GmbH: Frankfurt, Germany, 2011; Volume 8. (In German)
13. Martinez-Fernandez, A.; Soldado, A.; De-la-Roza-Delgado, B.; Vicente, F.; Gonzalez-Arrojo, M.A.; Argamenteria, A. Modelling a quantitative ensilability index adapted to forages from wet temperate areas. *Span. J. Agric. Res.* **2013**, *11*, 455–462. [CrossRef]
14. Verband deutscher landwirtschaftlicher Untersuchungs- und Forschungsanstalten (VDLUFA). *Handbuch der landwirtschaftlichen Versuchs- und Untersuchungsmethodik (VDLUFA-Methodenbuch). Band III. Die chemische Untersuchung von Futtermitteln*; VDLUFA-Verlag: Darmstadt, Germany, 2012; Volume 3. (In German)
15. Weissbach, F.; Kuhla, S. Substance losses in determining the dry matter content of silage and green fodder: Arising errors and possibilities of correction (Stoffverluste bei der Bestimmung des Trockenmassegehaltes von Silagen und Grünfutter: Entstehende Fehler und Möglichkeiten der Korrektur). *Übers. Tierernährg.* **1995**, *23*, 189–214. (In German)
16. Sniffen, C.J.; O'Connor, J.D.; van Soest, P.J.; Fox, D.G.; Russell, J.B. A net carbohydrate and protein system for evaluating cattle diets: II. Carbohydrate and protein availability. *J. Anim. Sci.* **1992**, *70*, 3562–3577. [CrossRef] [PubMed]
17. Licitra, G.; Hernandez, T.M.; van Soest, P.J. Standardization of procedures for nitrogen fractionation of ruminant feeds. *Anim. Feed Sci. Technol.* **1996**, *57*, 347–358. [CrossRef]
18. European Commission. Commission Regulation (EC) No. 152/2009 of 27 January 2009 laying down the methods of sampling and analysis for the official control of feed. *Off. J. Eur. Union* **2009**, *L54*, 1–130.

19. Steingaß, H.; Nibbe, D.; Südekum, K.-H.; Lebzien, P.; Spiekers, H. Estimation of uCP concentrations using the modified Hohenheim gas test and its application on the evaluation of rapeseed and soybean meals (Schätzung des nXP-Gehaltes mit Hilfe des modifizierten Hohenheimer Futterwerttests und dessen Anwendung zur Bewertung von Raps-und Sojaextraktionsschroten). In *Kurzfassungen der Vorträge, 113*; VDLUFA-Kongress: Berlin, Germany, 2001; 114p. (Abstr. in German).
20. Steingaß, H.; Südekum, K.H. Protein evaluation for ruminants—basics, analytical developments and perspectives (Proteinbewertung beim Wiederkäuer—Grundlagen, analytische Entwicklungen und Ausblick). *Übers. Tierernährg.* **2013**, *41*, 51–73. (In German)
21. Edmunds, B.; Südekum, K.-H.; Spiekers, H.; Schuster, M.; Schwarz, F.J. Estimating utilisable crude protein at the duodenum, a precursor to metabolisable protein for ruminants, from forages using a modified gas test. *Anim. Feed Sci. Technol.* **2012**, *175*, 106–113. [CrossRef]
22. Leberl, P.; Gruber, L.; Steingaß, H.; Schenkel, H. Comparison of the methods modified Hohenheimer Futterwerttest (moHFT) and Cornell system for determination of nXP-content of concentrates. In *Proceedings of the 16th International Science Symposium on Nutrition of Domestic Animals, Radenci, Slovenia, 8–9 November 2007*; Kapun, S., Kramberger, B., Ceh, T., Eds.; Kmetijsko gozdarska zbornica Slovenije, Kmetijsko gozdarski zavod: Murska Sobota, Slovenia, 2007; pp. 171–176.
23. Menke, K.-H.; Steingaß, H. Estimation of the energetic feed value obtained from chemical analysis and *in vitro* gas production using rumen fluid. *Anim. Res. Dev.* **1988**, *28*, 7–55.
24. Agricultural and Food Research Council. *Energy and Protein Requirements of Ruminants: An Advisory Manual Prepared by the AFRC Technical Committee on Responses to Nutrients*; CAB International: Wallingford, UK, 1993.
25. Brüning, D.; Gerlach, K.; Weiß, K.; Südekum, K.-H. Effect of compaction, delayed sealing and aerobic exposure on maize silage quality and on formation of volatile organic compounds. *Grass Forage Sci.* **2018**, *73*, 53–66. [CrossRef]
26. Hinds, A.A.; Lowe, L.E. Application of the Berthelot reaction to the determination of ammonium-N in soil extracts and soil digests. *Commun. Soil Sci. Plant Anal.* **1980**, *11*, 469–475. [CrossRef]
27. Weiß, K.; Kaiser, E. The determination of lactic acid in silage with HPLC (Milchsäurebestimmung in Silageextrakten mit Hilfe der HPLC). *Wirtschaftseig. Futter* **1995**, *41*, 69–80. (In German)
28. Weiß, K. Fermentation Process and Fermentation Quality of Silages Originating from Low-Nitrate Herbage (Gärungsverlauf und Gärqualität von Silagen aus nitratarmen Grünfutter). Ph.D. Thesis, Humboldt-Universität zu Berlin, Berlin, Germany, 2001. (In German)
29. Weiß, K.; Sommer, G. Determination of Esters and Other Volatile Organic Compounds (VOC) in Silage Extracts Using Gas Chromatography (Bestimmung von Estern und anderen flüchtigen organischen Substanzen (VOC) in Silageextrakten mit Hilfe der Gaschromatographie). *VDLUFA Schriftenr.* **2012**, *68*, 561–569. (In German)
30. Von Lengerken, J.; Zimmermann, K. *Handbuch Futtermittelprüfung*; Deutscher Landwirtschaftsverlag: Berlin, Germany, 1991; Volume 1. (In German)
31. Udén, P.; Robinson, P.H.; Mateos, G.G.; Blank, R. Use of replicates in statistical analyses in papers submitted for publication in Animal Feed Science and Technology. *Anim. Feed Sci. Technol.* **2012**, *171*, 1–5. [CrossRef]
32. Lowry, S.R. Use and misuse of multiple comparisons in animal experiments. *J. Anim. Sci.* **1992**, *70*, 1971–1977. [CrossRef] [PubMed]
33. Dewar, W.A.; McDonald, P.; Whittenbury, R. The hydrolysis of grass hemicelluloses during ensilage. *J. Sci. Food Agric.* **1963**, *14*, 411–417. [CrossRef]
34. Santos, M.C.; Kung, L. Short communication: The effects of dry matter and length of storage on the composition and nutritive value of alfalfa silage. *J. Dairy Sci.* **2016**, *99*, 5466–5469. [CrossRef] [PubMed]
35. Wyss, U.; Girard, M.; Grosse Brinkhaus, A.; Dohme-Meier, F. Crude protein fractions in three legume species (Proteinfraktionen in drei Leguminosenarten). *Agrarforsch. Schweiz* **2017**, *8*, 220–225. (In German)
36. Rotz, C.A. Field Curing of Forages. In *Post-harvest Physiology and Preservation of Forages*; CSSA Special Publication 22; Crop Science Society of America: Fitchburg, WI, USA; American Society of Agronomy: Madison, WI, USA, 1995; pp. 39–66.
37. Kirchhof, S.; Eisner, I.; Gierus, M.; Südekum, K.-H. Variation in the contents of crude protein fractions of different forage legumes during the spring growth. *Grass Forage Sci.* **2010**, *65*, 376–382. [CrossRef]

38. Yuan, X.; Wen, A.; Desta, S.T.; Dong, Z.; Shao, T. Effects of four short-chain fatty acids or salts on the dynamics of nitrogen transformations and intrinsic protease activity of alfalfa silage. *J. Sci. Food Agric.* **2017**, *97*, 2759–2766. [CrossRef]

39. Heron, S.J.E.; Edwards, R.A.; Phillips, P. Effect of pH on the activity of ryegrass *Lolium multiflorum* proteases. *J. Sci. Food Agric.* **1989**, *46*, 267–277. [CrossRef]

40. Tao, L.; Guo, X.S.; Zhou, H.; Undersander, D.J.; Nandety, A. Short communication: Characteristics of proteolytic activities of endo- and exopeptidases in alfalfa herbage and their implications for proteolysis in silage. *J. Dairy Sci.* **2012**, *95*, 4591–4595. [CrossRef]

41. Owens, V.N.; Albrecht, K.A.; Muck, R.E. Protein degradation and ensiling characteristics of red clover and alfalfa wilted under varying levels of shade. *Can. J. Plant Sci.* **1999**, *79*, 209–222. [CrossRef]

42. Muck, R.E. Dry matter level effects on alfalfa silage quality I. Nitrogen transformations. *Transact. ASAE* **1987**, *30*, 7–14. [CrossRef]

43. Purwin, C.; Sienkiewicz, S.; Pysera, B.; Lipiński, K.; Fijałkowska, M.; Piwczyński, D.; Puzio, N. Nitrogen fractions and amino acid content in alfalfa and red clover immediately after cutting and after wilting in the field. *J. Elementol.* **2014**, *19*, 723–734. [CrossRef]

44. Scherer, R.; Gerlach, K.; Südekum, K.-H. Biogenic amines and gamma-amino butyric acid in silages: Formation, occurrence and influence on dry matter intake and ruminant production. *Anim. Feed Sci. Technol.* **2015**, *210*, 1–16. [CrossRef]

45. Ohshima, M.; McDonald, P. A review of the changes in nitrogenous compounds of herbage during ensilage. *J. Sci. Food Agric.* **1978**, *29*, 497–505. [CrossRef]

46. Oh, C.-H.; Oh, S.-H. Effects of germinated brown rice extracts with enhanced levels of GABA on cancer cell proliferation and apoptosis. *J. Med. Food* **2004**, *7*, 19–23. [CrossRef]

47. Krizsan, S.J.; Westad, F.; Adnøy, T.; Odden, E.; Aakre, S.E.; Randby, A.T. Effect of volatile compounds in grass silage on voluntary intake by growing cattle. *Animal* **2007**, *1*, 283–292. [CrossRef]

48. Aschenbach, J.R.; Gäbel, G. Effect and absorption of histamine in sheep rumen: Significance of acidotic epithelial damage. *J. Anim. Sci.* **2000**, *78*, 464. [CrossRef]

49. Repetto, J.L.; González, J.; Cajarville, C. Effect of dehydration on ruminal degradability of lucerne. *Anim. Res.* **2000**, *49*, 113–118. [CrossRef]

50. Kung, L.; Shaver, R. Interpretation and use of silage fermentation analysis reports. *Focus Forage* **2001**, *3*, 1–5.

51. Driehuis, F.; Oude Elferink, S.J.W.H.; Spoelstra, S.F. Anaerobic lactic acid degradation during ensilage of whole crop maize inoculated with *Lactobacillus buchneri* inhibits yeast growth and improves aerobic stability. *J. Appl. Microbiol.* **1999**, *87*, 583–594. [CrossRef] [PubMed]

52. Weiß, K.; Kalzendorf, C. Effect of wilting and silage additives on silage quality of lucerne, red clover and legume-grass mixtures. In Proceedings of the 26th General Meeting of the European Grassland Federation, Trondheim, Norway, 4–8 September 2016; pp. 170–172.

53. Wylam, C.B. Analytical studies on the carbohydrates of grasses and clovers. III.—Carbohydrate breakdown during wilting and ensilage. *J. Sci Food Agric.* **1953**, *4*, 527–531. [CrossRef]

54. Zheng, M.; Niu, D.; Zuo, S.; Mao, P.; Meng, L.; Xu, C. The effect of cultivar, wilting and storage period on fermentation and the clostridial community of alfalfa silage. *Ital. J. Anim. Sci.* **2017**, *17*, 336–346. [CrossRef]

55. Đorđević, N.Ž.; Grubić, G.A.; Stojanović, B.D.; Božičković, A.Đ. The influence of compression level and inoculation on biochemical changes in lucerne silages. *J. Agric. Sci.* **2011**, *56*, 15–23. [CrossRef]

56. Eisner, I.; Südekum, K.-H.; Kirchhof, S. Relationships between silage fermentation characteristics and feed intake by dairy cows (Beziehungen zwischen Fermentationscharakteristika von Silagen und der Futteraufnahme von Milchkühen). *Übers. Tierernährg.* **2006**, *34*, 197–221. (In German)

57. Davies, D.R.; Merry, R.J.; Williams, A.P.; Bakewell, E.L.; Leemans, D.K.; Tweed, J.K. Proteolysis during ensilage of forages varying in soluble sugar content. *J. Dairy Sci.* **1998**, *81*, 444–453. [CrossRef]

58. Danner, H.; Holzer, M.; Mayrhuber, E.; Braun, R. Acetic acid increases stability of silage under aerobic conditions. *Appl. Environ. Microbiol.* **2003**, *69*, 562–567. [CrossRef]

59. Weiß, K.; Kroschewski, B.; Auerbach, H. Effects of air exposure, temperature and additives on fermentation characteristics, yeast count, aerobic stability and volatile organic compounds in corn silage. *J. Dairy Sci.* **2016**, *99*, 8053–8069. [CrossRef]

60. Pahlow, G.; Muck, R.E.; Driehuis, F.; Oude Elferink, S.J.W.H.; Spoelstra, S.F. Microbiology of Ensiling. In *Silage Science and Technology*; Buxton, D.R., Muck, R.E., Harrison, J.H., Eds.; American Society of Agronomy: Madison, WI, USA; Crop Science Society of America: Fitchburg, WI, USA; Soil Science Society of America: Madison, WI, USA, 2003; pp. 31–93.

61. Gerlach, K.; Roß, F.; Weiß, K.; Büscher, W.; Südekum, K.-H. Changes in maize silage fermentation products during aerobic deterioration and effects on dry matter intake by goats. *Agric. Food Sci.* **2013**, *22*, 168–181. [CrossRef]

62. Lorenzo, B.F.; O'Kiely, P. Alternatives to formic acid as a grass silage additive under two contrasting ensilability conditions. *Irish J. Agric. Food Res.* **2008**, *47*, 135–149.

63. Dawson, L.E.R.; Ferris, C.P.; Steen, R.W.J.; Gordon, F.J.; Kilpatrick, D.J. The effects of wilting grass before ensiling on silage intake. *Grass Forage Sci.* **1999**, *54*, 237–247. [CrossRef]

64. Savoie, P. Intensive mechanical conditioning of forages: A review. *Can. Biosyst. Eng.* **2001**, *43*, 2.1–2.12.

65. Mandell, I.B.; Mowat, D.N.; Bilanski, W.K.; Rai, S.N. Effect of heat treatment of alfalfa prior to ensiling on nitrogen solubility and in vitro ammonia production. *J. Dairy Sci.* **1989**, *72*, 2046–2054. [CrossRef]

agriculture

MDPI

Article

Sowing Date Affects Maize Development and Yield in Irrigated Mediterranean Environments

Angel Maresma, Astrid Ballesta, Francisca Santiveri and Jaume Lloveras *

Agrotecnio Center, University of Lleida, Rovira Roure 191, 25198 Lleida, Spain;
angel.maresma@pvcf.udl.cat (A.M.); astrid@hbj.udl.es (A.B.); santiveri@pvcf.udl.cat (F.S.)
* Correspondence: Jaume.lloveras@udl.cat

Received: 3 February 2019; Accepted: 21 March 2019; Published: 26 March 2019

Abstract: Timely sowing is critical for maximizing yield for both grain and biomass in maize. The effects of early (mid-March), normal (mid-April), and late (mid-May) sowing date (SD) were studied over a three-year period in irrigated maize under Mediterranean conditions. Early SD increased the number of days from sowing to plant emergence. Late SD reduced the number of days to plant maturity, and had higher forage yields, higher grain humidity, and taller plants. The average grain and forage yields achieved were 13.2 and 21.3 Mg ha^{-1}; 14.0 and 25.1 Mg ha^{-1}; and 12.8 and 27.6 Mg ha^{-1}, for crops with early, normal, and late SD, respectively. The data support the general perception of farmers that April sowings are the most appropriate in the area where the experiments were carried out. Early SD resulted in lower population densities, while later SD did not yield (grain) as high. However, late SD produced taller plants that contributed to achieve higher forage yields. Late SD could be interesting for double annual forage cropping systems. Sowing at the most appropriate time, when the soil is warm, ensures a good level of maize grain production. Future research could focus in the effect of SD for total annual yields in double-annual cropping systems.

Keywords: corn; forage yield; grain yield; plant height; planting; population density; sowing date

---

Highlights: Sowing date affects the average maize grain and forage yields. Germination and population density was reduced in mid-March sowing date. Traditional sowing date (mid-April) achieved highest grain yields. Mid-May sowing date was the most appropriate for forage production.

1. Introduction

Timely sowing is critical for maximizing yield for both grain and biomass in maize [1,2] and therefore, growers are concerned about the yield response of maize to sowing date (SD) [3–5]. However, optimum maize SD may vary from area to area due to differences in climate and the length of the growing season where the crop is produced [6].

It is known that maize needs warm soil to germinate and grow [5,7]. However, the practice of sowing as soon as possible to take advantage of the solar radiation [8] is nowadays more adopted by farmers. Early planting could contribute to the profitability of maize by increasing yields (crop has more time to photosynthesize) and, in some areas, by avoiding artificial grain drying at the end of the cycle [9–11].

Breeding programs have facilitated germination of maize at colder temperatures [12,13]. Bruns and Abbas [6] reported technological improvements in maize hybrids such as better early season vigor and tolerance to germination in cool wet soils, better seed treatments to guard against damping off diseases and seedling insect pests, or the advent of herbicides. These factors have contributed to planting maize earlier than it was 30 years ago [5].

In general, early sowing is preferable, but temperatures must be high enough to ensure quick germination and emergence. Also, SD must be late enough to avoid late spring frosts. As a rule, maize should not be sown until the soil temperature approaches 10 °C. Under cold soil conditions (below

10 °C), seeds will readily absorb water but will not initiate root or shoot growth, which leads to seed rot and poor emergence [5,14].

Increases in temperature during the vegetative period of maize crops hastens the growth rate more than the development rate, resulting in taller plants with a larger biomass [15]. Thus, under field conditions, rising temperature reduces the duration of crop growth, and consequently SD reduces the time during which incident radiation can be intercepted and transformed into dry matter (DM) [16].

The highest yields generally occur where the growing season is longest and soil moisture is not a limiting factor [17]. Yield reductions due to early or late planting have been well documented in the literature [4,5,9,11,16,18]. Early planting results in reduced cumulative intercepted photosynthetically active radiation (IPAR) because of delayed leaf area development. High temperatures under late planting scenarios also reduce cumulative IPAR by reducing the calendar time for crop development, and thereby, decreasing yields [19].

Optimum SD vary from one environment to another [8]. In the Ebro valley, the month of April is the most recommended sowing period for maize, particularly, the first half of the month [20–22]. Even so, there can be year-to-year variations associated with temperatures and rainfall during spring. In Mediterranean areas, maize is often grown under irrigation. Thus, crop growth depends on water availability during the growing season, and in some areas on irrigation turns. Many farmers currently practice double-annual cropping of maize after winter forage in order to increase the economic viability of their farms [23]. Therefore, in Mediterranean environments, a large variability exists in maize planting date, and there is a need to quantify the effect of the SD on maize yield.

The objective of this research was to evaluate the effects of the date of sowing (early, normal, and late) on maize yields (grain and forage) and crop growing period, in irrigated Mediterranean environments.

2. Materials and Methods

A three-year experiment (2003–2005) was conducted at the IRTA experimental station at Gimenells, Catalonia, Spain (41°65' N, 0°39' E), under sprinkler irrigated conditions. The study area is characterized by a semi-arid climate with low annual precipitation (345 mm) and a high annual average temperature (14.6 °C) (Table 1).

Table 1. Mean monthly (T_m) air temperatures and total monthly rainfall at Gimenells, during the experiment (from 2003 to 2005). Long-term (30 year) mean annual temperature and rainfall values at Gimenells are 14.6 °C and 345 mm, respectively.

Month	2003		2004		2005	
	T_m	Rainfall	T_m	Rainfall	T_m	Rainfall
	(°C)	(mm)	(°C)	(mm)	(°C)	(mm)
February	5.7	70.6	4.7	50.6	4.1	8.6
March	10.9	30.5	8.3	37.4	9.3	9.2
April	13.4	25.9	11.6	61.1	13.7	7.2
May	17.6	62.5	15.8	40.8	18.4	53.1
June	24.9	15.1	22.6	7.6	23.2	13.1
July	25.1	2.0	23.1	49.8	24.5	18.5
August	25.9	38.0	23.7	6.0	22.4	20.9
September	19.3	101.5	20.6	18.5	19.3	28.6
October	13.9	71.2	16.1	31.0	15.8	82.0

The soil was a Petrocalcic Calcixerept [24], which is representative of many areas of the Ebro valley (Table 2). Two maize cultivars with different growth cycles were sown: Cecilia (600 FAO) and Eleonora (700 FAO). The cultivars provided good representations of the 600 to 700 FAO cycles, which are the ones most commonly used in the area [20].

Table 2. Soil properties at the beginning of the experiment (2003).

Soil Properties	Horizon		
	Ap 0–25 cm	Bwk$_1$ 25–70 cm	Bwk$_2$ 70–120 cm
Sand, %	38.5	38.4	44.6
Silt, %	40.3	41.9	38.4
Clay, %	21.2	19.7	17.0
pH	8.1	8.2	8.3
Organic matter, g kg^{-1}	22.0	14.0	6.2
EC$_{1:5}$, dS m^{-1}	0.20	0.34	0.59
N (N-NO$_3$$^-$), mg kg^{-1}	33	-	-
P (Olsen), mg kg^{-1}	38	20	10
K (NH$_4$Ac), mg kg^{-1}	241	94	59

The statistical design of the maize experiments was a split plot, with four replications, where SD were used as the main plot, and cultivars as the subplots [25]. For each year, the different plots (harvest date) and subplots (cultivars) were randomly distributed.

Maize was sown at three different dates, which were as close as possible to 15 March, 15 April, and 15 May. The exact dates are presented in Table 3.

Table 3. Dates of sowing and harvesting for maize. The sowing rate was 85,000 plants ha^{-1} with a distance of 71 cm between rows. The plot size of each experimental plot was of 15 × 11 m.

Year	Sowing Date			Harvest Date	
	1	2	3	Biomass	Grain
2003	27 March	14 April	14 May	16–29 September	1 October
2004	15 March	14 April	17 May	8–22 September	5 October
2005	14 March	14 April	19 May	12 September	10 October

Conventional tillage was carried out before sowing. This included disc ploughing and cultivation to a depth of 30 cm to incorporate previous stover and to prepare the soil for the sowing. The maize was fertilized with 50 kg N ha^{-1}, 150 kg ha^{-1} of P$_2$O$_5$ and 200 kg ha^{-1} K$_2$O before sowing, and then two equal side dressings of 100 kg N ha^{-1} were applied at V4–V5 and at V6–V7 maize growing stages. Nitrogen was applied as ammonium nitrate (34.5% N), and maize was irrigated after the application to avoid N losses. In each growing season, around 650 mm of irrigation water were applied to the maize crop. Irrigation water was of good quality and did not contain any significant amounts of nitrates.

A pre-emergence herbicide (1 L ha^{-1} 96% metolachlor and 3 L ha^{-1} 47.5% atrazine) was applied to control weeds. When necessary hand wedding or post-emegence herbicide were applied to control *Abutilon theophrasti* (Banvel (20% fluoxypyr) at a rate of 1 L ha^{-1}), and to control Sorghum *halepense* (Elite (nicosulfuron 4%) at a rate of 1.5 kg ha^{-1}).

The height of 10 plants of the central rows was measured about one week after silking in each plot from the base of the crop to the last leaf. At same time, the leaf are index (LAI) was measured by taking all the leaves of five consecutive plants from one central row and measuring them with the LAI meter (Li-Cor, Lincoln, NE, USA). Intercepted solar radiation was measured by taking eight readings per plot from the central rows, at noon, using a Ceptometer (Delta-T devices, Burnell, UK). The plant density was estimated before forage harvest, counting the total plants of the two central rows in 5 m strips (1.42 m by 5 m).

The forage and grain harvest took place during September or October, after the plants had reached physiological maturity (Table 3). Grain yield was measured by harvesting two central rows (1.42 m by 15 m) from each plot using an experimental plot combine. Grain moisture was determined from a 300-g grain sample taken from each plot, using a GAC II (Dickey-John, Auburn, IL, USA), and the grain yield was adjusted to 14% moisture. The aboveground biomass yield was determined at physiological maturity by harvesting plants from one central row (0.71 m ɔy 5 m) at ground level. Subsamples were chopped and dried in a stove at 65 °C for at least 48 h to determine the DM weight.

A mixed-effects analysis of variance (ANOVA) was carried out to assess the responses to SD, with years evaluated as repeated measurements.

3. Results and Discussion

3.1. Duration of the Growing Cycle

The duration of the growing cycle (number of days from sowing to physiological maturity) decreased with each delay in sowing, falling from an average of 162 days with the earliest sowings (mid-March), to 143 and 125 days with the sowings on mid-April and on mid-May (Table 4), respectively.

The average number of days from sowing to plant emergence in the mid-March SD was 22, and was reduced to 12 and 9 days for the mid-April and mid-May SD, respectively. However, the greatest effect of delaying sowing was observed in the number of days from sowing to silking, which fell significantly, from 104 days in the mid-March SD, to 81 days (22% reduction) and to 69 days (33% reduction) for the mid-April and mid-May SD (Table 4). This is in agreement with Mederski and Jones [26], who reported a decrease in the number of days from sowing to silking as soil temperature increases. Moreover, the number of days from emergence to silking was considerably reduced in our study. It ranged from 82 to 60 days from the first to the last SD. There was a cultivar effect in the length of the growing cycle and the time to silking (Table 4). However, the time from sowing to emergence was similar in both cultivars. Eleonora (700 FAO) required 87 and 147 days from sowing to silk and to physiological maturity, respectively. Cecilia (600 FAO) required on average 4 and 6 days less than Eleonora, which represent about a 4% of the total time to arrive to each growing stage. These findings could be expected due to the different growing cycle of the cultivars.

Soils and air temperature were the main reason for these differences in growth duration. Warmer temperatures accelerate the rate of crop development, resulting in shorter vegetative and reproductive phases [26,27]. Although these differed from year to year because of annual variations in temperature (Table 1), growth duration clearly decreased when sowing was delayed. The average temperature in March only reached 10.9 °C in the first year of the experiment, whereas in the last two years it barely reached 9 °C. In the second year the average temperature in March was 8.3 °C, which is considered low for maize sowings and which occasioned a large period from sowing to emergence (28 days). Average April temperatures were not very high either. They were above 10 °C, but never exceeded 13.7 °C. During June, July and August, the average monthly temperatures ranged from 22.4 °C to 25.9 °C (Table 1).

Table 4. Sowing date (SD) (mid-March, mid-April, and mid-May) effect on average forage maize dry matter (DM) yield at maturity, grain yield and humidity, the dates of emergence, silk and black layer appearance, and the days happened between sowing and silk and black layer appearance, plant height, leaf area index (LAI) and intercepted solar radiation (measured one week after silking). Average 2003–2005.

Sowing Date (SD)	Forage Yield	Grain Yield (14% hum.)	Harvest Index	Grain Humidity	Emergence	Silk	Black Layer	Plant Height	Plant Density	LAI	Intercepted Solar Radiation
	(Mg ha^{-1})	(Mg ha^{-1})		(%)	(days)	(days)	(days)	(m)	(plants ha^{-1})	(m^2 m^{-2})	(%)
Mid-March	**21.3**	**13.2**	**0.60**	**15.7**	**22**	**104**	**162**	**2.21**	**70,764**	**3.57**	**84.9**
2003	22.0	12.5	0.57	16.5	20	93	144	2.25	73,541	4.30	82.8
2004	22.4	14.2	0.63	16.9	28	111	178	2.15	80,000	3.02	89.1
2005	19.6	13.0	0.66	13.9	18	108	164	2.24	78,333	3.39	82.7
Mid-April	**25.1**	**14.0**	**0.55**	**16.8**	**12**	**81**	**143**	**2.33**	**78,333**	**4.02**	**88.7**
2003	25.4	13.3	0.52	17.8	10	78	140	2.34	70,000	4.20	87.9
2004	27.8	15.7	0.56	17.7	14	84	152	2.41	80,833	4.12	95.1
2005	22.0	12.9	0.59	15.0	12	82	137	2.25	75,833	3.74	83.2
Mid-May	**27.6**	**12.8**	**0.45**	**24.9**	**9**	**69**	**125**	**2.53**	**75,069**	**4.88**	**91.9**
2003	27.6	11.5	0.42	27.5	9	68	126	2.61	68,750	5.50	96.7
2004	30.7	15.4	0.50	25.0	10	72	129	2.59	74,166	4.88	94.9
2005	24.5	11.5	0.47	22.4	8	66	121	2.38	71,042	4.26	84.0
					ANOVA						
Year (Y)	*	**	*	**	**	**	**	**	**	**	**
Error											
Sowing Date (SD)	**	**	**	**	**	**	**	*	**	**	**
SD*Y	*	*	*	**	**	**	**	**	*	ns	*
Error											
Cultivar (C)	ns	ns	ns	**	ns	**	**	*	ns	**	**
C*SD	ns	*	ns	**	ns	*	*	ns	ns	ns	ns
Y*C	ns	ns	ns	ns	ns	ns	**	ns	ns	*	*
Y*C*SD	ns	ns	ns	ns	ns	**	**	ns	ns	ns	ns

* Significant at *p*-value < 0.05; ** Significant at *p*-value < 0.01; ns = not significant.

3.2. Plant Height, LAI, Intercepted Solar Radiation, and Plant Density

As previously described by other authors [1,5,28], plant height increases with delayed sowing. In the present experiment, maize plants height increased significantly (13%) from an average of 2.21 m for the mid-March SD to 2.53 m for plants sown on May (Table 4). Warm weather during early vegetative growth can stimulate plants to develop larger vegetative structures [16]. Despite cultivars showed differences in plant height, in the three-year average plant height was 2.40 m for Eleonora and 2.32 m for Cecilia. Those differences in plant height were smaller than the observed between the SD. There were significant differences in the LAI. The taller plants of Eleonora contributed to obtain higher LAI and intercepted solar radiation than Cecilia. The LAI and the intercepted solar radiation were respectively 0.4 m^2 m^{-2} and 3% higher for Eleonora. Late SD averaged LAI values of 4.88 m^2 m^{-2}, whereas the SD of mid-March and mid-April averaged respectively 3.57 and 4.02 m^2 m^{-2}. Thus, the tallest plants were also the ones that obtained the highest LAI values. Consequently, the amount of intercepted solar radiation was also higher for the late SD (Table 4). The LAI indexes obtained in our trials were lower than those reported by Tsimba et al. [4] in New Zealand. However, the grain yields achieved were similar.

The amount of intercepted solar radiation differed depending on the growth periods associated with the different sowings. Probably, this fact could help to explain the differences in yield and DM content in the three SD (Table 4). As reported by Cirilo and Andrade [16], late sowings resulted in high crop growth rates during the vegetative period because of high radiation use efficiency (RUE) and high percent radiation interception. However, these treatments resulted in low crop growth rates during grain filling because of low RUE and low incident radiation.

March temperatures in the second and third growing seasons were below 10 °C (Table 1), which affected the germination of the maize in the early SD. In the second growing season, there were 28 days between sowing and emergence for the mid-March SD. This was a long period and some of the plants did not emerge at all. Only 70,000 of the 85,000 plants ha^{-1} initially sown emerged (with plant losses of about 17%). Therefore, poor maize germination is one of the possible consequences of early sowing in some years. Moreover, the slow growth of maize made it less competitive with weeds and more weed control was necessary. Other researchers [14] have also reported these kinds of results. Earlier sowing may not be the most interesting option, whereas sowing at the appropriate time when the soil is warm, tends to ensure a good plant stand. Both, year and SD were significant for the total plant density. Temperature plays an important role for the successful development of seed to plants, but others factors such as soil preparation or precipitation can influence the germination. The lowest plant density was determined for 2005, which had the driest early season for the studied fields. The final three-year average densities for the three SD were around 71,000 plants ha^{-1}, 78,000 plants ha^{-1} and 75,000 plants ha^{-1}, for the early, middle, and late SD, respectively.

The differences in LAI observed seem to confirm the need to increase the sowing density at early SD in order to achieve sufficient photosynthetically active radiation interception [4]. As previously mentioned, the number of days from sowing to silking decreases with increases in soil and air temperature. Indeed, temperature has a major influence on the rate of maize development [4,7,16,29]. According to Duncan [7], the rate of maize development, from sowing to anthesis, is a function of temperature more than of photosynthesis.

Photoperiod can also influence maize development and grain maturation [30,31]. However, the differences in photoperiod associated with the sowing periods considered in this study were not sufficiently large at the critical photoperiod-sensitive interval (at tassel initiation or at between stages V5 and V7) to have had an impact on plant development [30]. Past studies indicate that differences in photoperiod from 3 h to 5 h are needed during the photoperiod-sensitive interval to generate differences in the phenological response of the Corn Belt germplasm [31].

On the first year of the experiment, there were storms and strong winds few days before harvesting, and as a result, many plants were lodged. Around the 21–23% of the plants from the mid-March, and mid-April sowings were lodged, whereas the 74% of the plants from the last sowing (mid-May) were affected. The difference among SD could probably be affected by the height of the plants. The more optimal growing conditions (mainly temperature) for the last SD contributed to increase the height of the plant and made them more vulnerable to lodging when storms occur at the end of the crop cycle. In the other years, lodging was inconsequential.

3.3. Grain and Biomass Yields

The optimum SD for grain and biomass were similar in all of the studied years despite of some year-to-year variation (data not shown). Maize sown in mid-April achieved the highest average grain yields (14.0 Mg ha^{-1}), followed by mid-March sowings (13.2 Mg ha^{-1}) and the lowest grain yields were achieved with mid-May sowings (12.8 Mg ha^{-1}). Every year, mid-April SD yielded higher than the mid-March and mid-May alternatives, except for the last year where the mid-March SD achieved similar yields. That yield variability among SD (Table 4) was expected because of the different weather conditions of the experiment each year (year and SD*year were significant). Mid-April is the most common SD in the studied area, possibly because of the cooler temperatures during mid-March SD and the shorter growing season associated with mid-May SD. The average grain yield obtained in the study is similar to the averages reported by Cela et al. [32] (13.6 ± 0.4 Mg ha^{-1}) in the same area.

Grain humidity at harvest time increased with delays in sowing, varying from 24.9% at the last SD (mid-May) to 16.8% and 15.7% to the mid-April and mid-March SD, respectively. Cultivar also had an effect in grain humidity. The longer growing cycle of Eleonora (700 FAO) was translated into higher grain humidity at harvest (20.8%) compared with Cecilia (17.5%). This may prove important in a few years, because the drying of the grain increases the production costs and consequently the maize profitability.

Forage yield increased when delaying the SD, similarly to the results reported by Bunting [1], Dillon and Gwin [28], and Fairey [33]. Indeed, Mederski and Jones [26] reported that increasing soil temperature accelerates the rate of development of maize and produce significant increases in DM production. Although, in some conditions there were reported no differences when delaying the SD [34].

The highest biomass yields (27.6 Mg ha^{-1}) were associated with the latest sowings (May): in which the plants grew taller, although with less grain proportion than the earlier SD. The harvest index decreased from 0.6, in early sowings, to 0.54 in mid-period sowings and to 0.44 for late sowings (Table 4). This fact may be interesting for forage production farmers who use double cropping systems. They can grow a forage crop during winter and thereafter plant maize in summer, which have higher forage yield potential than growing a monocropped maize [23]. However, a quality analysis of the forages may be required as Deinum and Struik [35] and Bunting [1] suggested that delaying sowing may reduce forage digestibility because of the lower grain proportion.

Cirilo and Andrade [16] reported that crop DM partitioning was strongly affected by SD. Early sowing favored reproductive growth, whereas late sowing favored vegetative growth. Delays in the SD hastened plant development between seedling emergence and silking, reducing crop exposure to cumulative incident radiation during the vegetative period. Dobben [15] indicated that increases in temperature during the maize vegetative period hastened growth rate more than development rate, resulting in taller plants with larger biomasses.

Sowing date has a significant effect on maize grain yield when all other factors are equal. Research across the USA has shown that there is an 'ideal' sowing window, with a decline in grain yield with each additional day after it, as less light and fewer growing degree days are available to the plant [5,6,11]. However, this 'ideal' sowing window is not constant over the years and may vary according to the weather, as it happened in one out of three years of the study (third year), in which although without significant differences, the first SD obtained slightly higher grain yields (Table 4).

The influence of SD and plant density on maize grain yield may be related to IPAR and LAI. In Argentina, delayed SD have been shown to increase IPAR levels at the silking stage by increasing leaf area development as a result of higher temperatures during vegetative growth [16]. Even so, yields were still lower with delayed SD due to reduced levels of cumulative IPAR.

4. Conclusions

The duration of the growing cycle (number of days from sowing to physiological maturity) was reduced by each delay in sowing. Early sowings increased the period from sowing to plant emergence, which reduced the germination and the population density of the crop. Alternatively, late sowings reduced the number of days to physiological maturity producing higher humidity content in grain at harvest and taller plants.

Sowing dates of middle April seem to be the most interesting for achieving the maximum grain yields under irrigated Mediterranean conditions. To anticipate the SD may not always be interesting; it will depend on soil temperature of the year to ensure germination. However, if the interest is on the forage yields (albeit with a lower proportion of grain yield), sowing can be delayed in order to benefit from the higher growth of the maize at higher temperatures, as well as to open the window to grow a winter crop.

Author Contributions: Á.M. contributed in the analysis of the data and in the writing of the bulk of the paper. A.B. and F.S. acquired field data. J.L. conceived and designed the experiment, and contributed in the analysis of the data and in the writing of the bulk of the paper.

Funding: This work was supported by the DARP (Department of Agriculture, Livestock, Fishing and Food) of the Generalitat de Catalunya.

Acknowledgments: The authors would like to thank the students and technicians, Josep Pons, Pau Marcé, Albert Casals, Silvia Martí form UdL and Josep Anton Betbesé and Jose Luis Millera of the IRTA for their help with the sowing and harvesting of the crops.

Conflicts of Interest: The authors declare no conflict of interest. The funding sponsors had no role in the design of the study; in the collection, analyses, or interpretation of data; in the writing of the manuscript, or in the decision to publish the results.

Abbreviations

ADF	acid detergent fiber
CP	crude protein
DM	dry matter
LAI	leaf area index
NDF	neutral detergent fiber
SD	sowing date

References

1. Bunting, E.S. The influence of date of sowing on development and yield of maize in England. *J. Agric. Sci.* **1968**, *71*, 117–125. [CrossRef]
2. Van Roekel, R.J.; Coulter, J.A. Agronomic responses of corn hybrids to row width and plant density. *Agron. J.* **2012**, *104*, 612–620. [CrossRef]
3. Olson, R.A.; Sander, D.H. Corn production. In *Corn and Corn Improvement*; American Society of Agronomy: Madison, WI, USA, 1988.
4. Tsimba, R.; Edmeades, G.O.; Millner, J.P.; Kemp, P.D. The effect of planting date on maize grain yields and yield components. *Field Crop. Res.* **2013**, *150*, 135–144. [CrossRef]
5. Abendroth, L.J.; Woli, K.P.; Myers, A.J.W.; Elmore, R.W. Yield-based corn planting date recommendation windows for Iowa. *Crop. Forage Turfgrass Manag.* **2017**, *3*, 1–7. [CrossRef]
6. Bruns, H.A.; Abbas, H.K. Planting date effects on Bt and non-Bt corn in the mid-south USA. *Agron. J.* **2006**, *98*, 100–106. [CrossRef]
7. Duncan, W.G. Maize. In *Crop Physiology*; Evans, L.T., Ed.; Cambridge University Press: Cambridge, UK, 1975.
8. Andrade, F.H.; Cirilo, A.G.; Uhart, S.A.; Otegui, M.E. *Ecofisiología del Cultivo de Maíz (No. 633.15 584.92041)*; Dekalb Press: Balcarce, Argentina, 1996.

9. Johnson, R.R.; Mulvaney, D.L. Development of a model for use in maize replant decisions 1. *Agron. J.* **1980**, *72*, 459–464. [CrossRef]

10. Nafziger, E. *Corn. Illinois Agron. Handbook*; University of Illinois Extension: Urbana, IL, USA, 2009; Chapter 2; pp. 13–26.

11. Lauer, J.G.; Carter, P.R.; Wood, T.M.; Diezel, G.; Wiersma, D.W.; Rand, R.E.; Mlynarek, M.J. Corn hybrid response to planting date in the northern corn belt contribution. Univ. of Wisconsin dep. of agronomy. *Agron. J.* **1999**, *91*, 834–839. [CrossRef]

12. Pešev, N.V. Genetic factors affecting maize tolerance to low temperatures at emergence and germination. *Theor. Appl. Genet.* **1970**, *40*, 351–356. [CrossRef]

13. Sanghera, G.S.; Wani, S.H.; Hussain, W.; Singh, N.B. Engineering cold stress tolerance in crop plants. *Curr. Genom.* **2011**, *12*, 30–43. [CrossRef]

14. Hall, R.G.; Reitsma, K.D.; Clay, D.E. Best management practices for corn production in South Dakota: Corn planting guide. In *Grow Corn: Best Management Practices*; South Dakota State University: Brookings, SD, USA, 2016; Chapter 3; pp. 13–16.

15. Van Dobben, W.H. Influence of temperature and light conditions on dry-matter distribution, development rate and yield in arable crops. *Neth. J. Agric. Sci.* **1962**, *10*, 377–389.

16. Cirilo, A.G.; Andrade, F. Sowing Date and Maize Productivity: I. Crop growth and dry matter partitioning. *Crop Sci.* **1994**, *34*, 1039–1043. [CrossRef]

17. Kucharik, C.J. A multidecadal trend of earlier corn planting in the central USA. *Agron. J.* **2006**, *98*, 1544–1550. [CrossRef]

18. Nafziger, E.D. Corn planting date and plant population. *J. Prod. Agric.* **1994**, *7*, 59–62. [CrossRef]

19. Otegui, E.; Nicolini, G.; Ruiz, R.A.; Dodds, P.A. Sowing date effects on grain yield components for different maize genotypes. *Agron. J.* **1995**, *87*, 29–33. [CrossRef]

20. Sisquella, M.; Lloveras, J.; Álvaro, J.; Santiveri, P.; Cantero, C. *Técnicas de Cultivo para la Producción de Maíz, Trigo y Alfalfa en Regadíos del Valle del Ebro*; Fundació Catalana de Cooperació: Barcelona, Spain, 2004; ISBN 846687860X.

21. Martínez, E.; Maresma, A.; Biau, A.; Cela, S.; Berenguer, P.; Santiveri, F.; Michelena, A.; Lloveras, J. Long-term effects of mineral nitrogen fertilizer on irrigated maize and soil properties. *Agron. J.* **2017**, *109*, 1880–1890. [CrossRef]

22. Martínez, E.; Maresma, A.; Biau, A.; Berenguer, P.; Cela, S.; Santiveri, F.; Michelena, A.; Lloveras, J. Long-term effects of pig slurry combined with mineral nitrogen on maize in a Mediterranean irrigated environment. *Field Crop. Res.* **2017**, *214*, 341–349. [CrossRef]

23. Maresma, Á.; Martínez-Casasnovas, J.A.; Santiveri, F.; Lloveras, J. Nitrogen management in double-annual cropping system (barley-maize) under irrigated Mediterranean environments. *Eur. J. Agron.* **2019**, *103*, 98–107. [CrossRef]

24. *Soil Survey Staff Keys to Soil Taxonomy*; USDA-Natural Resources Conservation Service: Washington, DC, USA, 2014; Volume 12, ISBN 0926487221.

25. Steel, R.G.D.; Torrie, J.H. *Principles and Procedures of Statistics: A Biometrical Approach*; McGraw-Hill: New York, NY, USA, 1980.

26. Mederski, H.J.; Jones, J.B. Effect of soil temperature on corn plant development and yield: I. Studies with a corn hybrid. *Soil Sci. Soc. Am. J.* **1963**, *27*, 186–189. [CrossRef]

27. Lizaso, J.I.; Ruiz-Ramos, M.; Rodríguez, L.; Gabaldon-Leal, C.; Oliveira, J.A.; Lorite, I.J.; Sánchez, D.; García, E.; Rodríguez, A. Impact of high temperatures in maize: Phenology and yield components. *Field Crop. Res.* **2018**, *216*, 129–140. [CrossRef]

28. Dillon, M.A.; Gwin, R.E., Jr. *How Planting Date and Full-Season or Early Hybrids Affect Corn Yields*; Bulletin 600; Agricultural Experiment Station, Kansas State University: Manhattan, KS, USA, 1976.

29. Carr, M.K.V. The influence of temperature on the development and yield of maize in Britain. *Ann. Appl. Biol.* **1977**, *87*, 261–266. [CrossRef]

30. Kiniry, J.R.; Ritchie, J.T.; Musser, R.L.; Flint, E.P.; Iwig, W.C. The photoperiod sensitive interval in maize1. *Agron. J.* **1983**, *75*, 687. [CrossRef]

31. Tollenaar, M.; Hunter, R.B. A Photoperiod and temperature sensitive period for leaf number of maize. *Crop Sci.* **1983**, *23*, 457–460. [CrossRef]

32. Cela, S.; Berenguer, P.; Ballesta, A.; Santiveri, F.; Lloveras, J. Prediction of relative corn yield with soil-nitrate tests under irrigated mediterranean conditions. *Agron. J.* **2013**, *105*, 1101–1106. [CrossRef]

33. Fairey, N.A. Yield, quality and development of forage maize as influenced by dates of planting and harvesting. *Can. J. Plant Sci.* **1983**, *63*, 157–168. [CrossRef]

34. Opsi, F.; Fortina, R.; Borreani, G.; Tabacco, E.; López, S. Influence of cultivar, sowing date and maturity at harvest on yield, digestibility, rumen fermentation kinetics and estimated feeding value of maize silage. *J. Agric. Sci.* **2013**, *151*, 740–753. [CrossRef]

35. Deinum, B.; Struik, P.C.; Dolstra, O.; Miedema, P. *Breeding of Silage Maize; Proceedings of the 13th Congress of Maize and Sorghum Section of Eucarpia*; Pudoc: Wageningen, The Netherlands, 1986; ISBN 9022008959.

agriculture

MDPI

Article

Utilization of Molecular Marker Based Genetic Diversity Patterns in Hybrid Parents to Develop Better Forage Quality Multi-Cut Hybrids in Pearl Millet

Govintharaj Ponnaiah [1,2,*], Shashi Kumar Gupta [2], Michael Blümmel [3], Maheswaran Marappa [1], Sumathi Pichaikannu [1], Roma Rani Das [2] and Abhishek Rathore [2]

[1] Tamil Nadu Agricultural University, Coimbatore 641 003, Tamil Nadu, India; mahes@tnau.ac.in (M.M.); sumivetri@yahoo.com (S.P.)
[2] International Crops Research Institute for the Semi-Arid Tropics (ICRISAT), Patancheru 502 324, Hyderabad, Telangana, India; s.gupta@cgiar.org (S.K.G.); r.das@cgiar.org (R.R.D.); a.rathore@cgiar.org (A.R.)
[3] International Livestock Research Institute (ILRI), Patancheru 502 324, Hyderabad, Telangana, India; m.blummel@cgiar.org
[*] Correspondence: p.govintharaj@cgiar.org; Tel.: +91-9502586220

Received: 10 April 2019; Accepted: 24 April 2019; Published: 3 May 2019

Abstract: Genetic diversity of 130 forage-type hybrid parents of pearl millet was investigated based on multiple season data of morphological traits and two type of markers: SSRs (Simple sequence repeats) and GBS identified SNPs (Genotyping by sequencing-Single nucleotide polymorphism). Most of the seed and pollinator parents clustered into two clear-cut separate groups based on marker based genetic distance. Significant variations were found for forage related morphological traits at different cutting intervals (first and second cut) in hybrid parents. Across two cuts, crude protein (CP) varied from 11% to 15%, while *in vitro* organic matter digestibility (IVOMD) varied from 51% to 56%. Eighty hybrids evaluated in multi-location trial along with their parents for forage traits showed that significant heterosis can be realized for forage traits. A low but positive significant correlation found between SSR based genetic distance (GD between parents of hybrid) and heterosis for most of the forage traits indicated that SSR-based GD can be used for predicting heterosis for GFY, DFY and CP in pearl millet. An attempt was made to associate marker-based clusters with forage quality traits, to enable breeders select parents for crossing purposes in forage breeding programs.

Keywords: genetic diversity; markers; forage yield; crude protein; *in vitro* organic matter digestibility

---

1. Introduction

Pearl millet (*Pennisetum glaucum* (L.) R. Br.) is an important staple crop in the arid and semi-arid tropical regions of Asia and Africa. This crop, being a C_4 species, is highly photosynthetically efficient, has a short duration coupled with significant levels of pest and disease resistance, and is tolerant to abiotic stresses (drought, heat & salinity). As such, it can be designated as a "perfect resilient crop for the future". This crop is mainly cultivated for grain and fodder purpose in semi-arid regions of Asia and Africa, for which dual purpose cultivars are popular on farms. Apart from dual-purpose cultivars, pearl millet cultivars are also bred exclusively for forage purpose and cultivated across the globe. For instance, pearl millet forage hybrids are grown in southern USA [1,2] and in the summer season in Australia and South America [3]. Recently, it has occupied large areas under summer cultivation in north-western India and is proving to be significant source of fodder for livestock [4,5]. Also, Brazil which introduced this crop as cover crop in soybean cropping system, now cultivates 5 m ha of pearl millet for feed and forage purposes [6,7].

Non-availability of feed and fodder in sufficient amounts has been one of the major limiting components to achieving the desired level of livestock production in most of the countries in the semi-arid tropics. For instance, India faces a net deficit of 35.6% green fodder, 10.95% of dry crop residues and 44% of concentrate feeds, and it would require 1012 million tons of green fodder and 631 million tons of dry fodder by 2050. At the current level of growth in forage production, there will be a 18.4% deficit in green fodder and 13.2% deficit in dry fodder by 2050 in India [8]. To overcome this projected deficit, green forage supply should grow in India an the annual rate of 1.69%. Breeding for forage traits has not been prioritized in most of the pearl millet programs across the globe, leading to a reduced diversity for forage cultivars, with only a handful of cultivars being available for cultivation. The pearl millet forage hybrids released to date have been bred generally for single cut forage purposes, but farmers are now demanding multi-cut (2–3 cuts) forage cultivars with better forage quality to meet round the year feed requirements of livestock [9]. Forage trials of pearl millet conducted multi-locationally during summer season in India, reported 10% to 16% higher dry stover yields than sorghum and 21% to 30% higher stover yields than maize; 45% to 64% and 30% to 58% higher stover protein was reported than sorghum and maize, respectively [10]. Significant variability has been observed in previous studies for forage quality traits in pearl millet breeding materials [11]. Also, based on evaluation of pearl millet accessions for biomass traits, high variability was reported for forage traits like green forage yield (GFY), dry forage yield (DFY), stover nitrogen content, metabolizable energy (ME) and *in vitro* organic matter digestibility (IVOMD) [12].

To date, almost all the investigations on characterization of pearl millet (hybrid parents or other breeding materials) are either based on morphological traits or on molecular markers, and have primarily targeted grain yield and component traits [13–20]. A positive association has been reported between molecular marker-based GD and hybrid performance in pearl millet [20], while few studies have indicated a negative relationship [17,21]. Similarly, some studies found a positive relationship between molecular marker based genetic diversity and heterosis in other crops, like in maize [22,23], in rice [24,25] and in sunflower [26], while others reported no relationship between GD and heterosis in maize [27,28], in rice [29,30] and in sunflower [31]. Hence, the present study was designed to investigate genetic diversity in forage type hybrid parents based on both morphological traits and also using two different molecular marker systems (SSRs and GBS-identified SNPs markers), to reveal the relationship between marker based genetic diversity and heterosis for forage traits.

2. Materials and Methods

2.1. Plant Materials

A set of 130 hybrid parents derived from high biomass nursery (F_6 and above) at ICRISAT (International Crops Research Institute for the Semi-Arid Tropics)-Patancheru, Hyderabad, Telangana, India, was investigated in this study. This included 18 seed parents and 112 pollinator parents, and all of them were derived from crosses involving diverse parents following pedigree breeding (Table S1). The seed parents were coded from FB01 (Forage B line) to FB18, while pollinator parents were coded from FP01 (Forage pollinator) to FP112. Tift $23D_2B_1$, a maintainer of A_1 CMS (cytoplasmic male sterility) bred at Tifton, Georgia [32,33] was used as a reference genotype.

2.2. DNA Isolation

Around thirty five seeds of each entry were grown in small plastic (4 inch) pots along with Tift $23D_2B_1$ in a dark house for eight days. Approximately 100 mg of bulk leaf tissue was collected from 20 to 25 seedlings per accessions and stored immediately in a 96-well plate. DNA was isolated using NucleoSpin® 96 Plant II kit (Macherey-Nagel, Düren, Germany). Two elutions of DNA for SNP and SSR genotyping were generated. Normalization of genomic DNA to 10 ng/µL was done on 0.8% agarose gel using lamda DNA (MBI Fermentas, Hanover, MD, USA). Electrophoresis was performed in Tris acetate-EDTA buffer in a buffer tank at 90 volts for 1 h and gels were visualized by UV light using image analyzer after being stained with ethidium bromide.

2.3. Genotyping of Hybrid Parents

2.3.1. Simple Sequence Repeats

A total of 52 SSR markers (Table S2) were reported to be highly polymorphic, of which 47 were mapped earlier across 7 pearl millet linkage groups [34–38] and were used to genotype the set of forage type hybrid parents involved in this study.

2.3.2. Polymerase Chain Reaction

PCRs were performed using 10 μL volumes of reaction mixture, containing 2 μL of 10 ng DNA template, 0.5 μL of 1 mM dNTPs, and 0.06 μL of 0.2U Taq DNA polymerase, 1 μL of 10× Kappa Taq Polymerase buffer with MgCl₂, and 1 μL of primer containing 2 pmol/μL of forward and 4 pmol/μL of reverse primer, 0.2 μL of dye either Fam, Vic, Ned and Pet. PCR amplification was carried out using step-down program in a thermal cycler (GeneAmp, PCR System 9700; Applied Biosystems, Foster City, CA, USA) using 384-well PCR plates. The amplification conditions of initial denaturation at 94 °C for 5 min, 10 cycles at 94 °C for 25 s, 64 °C (−1 °C/cycles) for 20 s, and 72 °C for 30 s, followed by 37 cycles at 56 °C for 20 s, and 72 °C for 30 s, with a final extension at 72 °C for 20 min. After amplification, PCR products were multiplexed with 1 μL of each of dye-labeled products (Fam, Vic, Ned and Pet), 7 μL of Hi-Di™ Formamide (Applied Biosystems, Foster City, California, CA, USA), 0.1 μL of the LIZ-labeled (500[−250]) internal size standard, and 3.9 μL of distilled water. The DNA fragments were size separated on an ABI 3700 automatic DNA sequencer (Perkin-Elmer/Applied Biosystems, Foster City, CA, USA) using Gene Mapper® 4.0 software and GeneScan 3.1 (Applied Biosystems, Foster City, CA, USA) was used for allele calling. AlleloBin 2.0 [39] was used to measure the accurate allele size.

2.3.3. Genotyping by Sequencing (GBS) and SNP Calling

Hybrid parents were also genotyped using the GBS method as described by Elshire et al. [40]. The genomic libraries were constructed using an *ApeKI* endonuclease restriction enzyme. PCR amplification of pooled amplicons was carried out before sequencing on a Illumina Hiseq2500 platform (Illumina Inc., San Diego, CA, USA).

Raw sequencing reads and barcode information were processed with the non-reference based UNEAK (Universal Network Enabled Analysis Kit) pipeline [41] implemented in TASSEL V.4 software [42] to identify SNPs. Barcodes containing reads are retained and used for SNP calling. These reads are trimmed to 64 bp from barcode side, aligned against each other and used for SNP identification. Total of 19,652 SNPs were identified across all hybrid parents in pearl millet. Identified SNPs were assigned to each hybrid parents based on the barcode sequence information. Further, the SNP data were filtered with minor allele frequency (MAF) cutoff of 0.10 (10%) and SNP with ≥25% missing data. After filtering for missing data and minor allele frequency, we obtained 7870 SNPs to conduct diversity analysis on our set of hybrid parents.

2.4. Phenotyping of Hybrid Parents

A set of 116 hybrid parents was evaluated for forage yield and quality traits in partially balanced alpha lattice design with two replications, at ICRISAT, Patancheru (18° N, 78° E, 545 m above sea level) during the summer season of 2015–2016. The plots consisted of 4 rows of 4 m in length spaced at 60 cm.

During the time of field preparation, nitrogen and phosphorous were applied as a basal dose in the form of 100 kg ha⁻¹ of Diammonium phosphate (18% N and 46% P). Plots were fertilized equally with dosage rate of 100 kg ha⁻¹ of urea (46% N) as top dressing, two times before first harvest. Trial was irrigated at 12 to 15 days interval to avoid any moisture stress, and crop was protected from weeds, pests, animals and diseases. Green forage of each entry in the trial was first harvested at 50 days (around boot stage of plant development) after planting by cutting at the second node from the bottom of the plant. Fresh weight of the green forage was recorded (kg) for plot and converted into t ha^{-1}. A sub-sample (10–15 plants) of about 1 kg was collected per entry from the freshly harvested green forage and recorded for green forage weight, oven-dried for 8 h daily for three to four days at 60 °C in Campbell dryer (Campbell Industries, Inc., 3201 Dean Avenue, Des Moines, IA, USA), and weighed again (dry forage weight in kg). The dry matter concentration was determined by the ratio between the dry forage weight and green forage weight. DFY (t ha^{-1}) on a plot basis for each entry was calculated by multiplying the green forage weight and dry matter concentration. Dried sub-samples were chopped into 10 to 15 mm pieces using a chaff cutter (Model # 230, Jyoti Ltd., Vadodara, India) and ground using a Thomas Wiley mill (Model # 4, Philadephia, PA, USA) to pass through a 1-mm screen for chemical analysis. Ground stover samples (Approximately, 40 g of sample/entry) were analyzed by Near-Infrared Reflectance Spectroscopy (NIRS) for stover nitrogen concentration (N × 6.25 equals to CP content) and IVOMD [43,44]. A second cut of forage was taken after thirty days of the first cut; GFY, DFY and stover quality traits were again recorded as described in the first cut.

2.5. Hybrid Development and Phenotypic Evaluation

Ten seed and eight pollinator parents from the set of 130 hybrid parental lines were selected based on diverse pedigrees (Table S1). This set of 18 hybrid parents effectively represented the genetic diversity of all the 130 lines, as GD of selected lines varied from 0.42 to 0.79 with a mean of 0.64, while GD of all the lines varied from 0.38 to 0.95 with a mean of 0.74. For SNPs, GD varied from 0.22 to 0.58 with a mean of 0.46 for selected hybrid parents, while GD of whole set of hybrid parents varied from 0.18 to 0.64 with a mean of 0.47. Eighty hybrids were generated by crossing them in a line × tester (10 × 8 = 80) fashion. These 80 hybrids and their parents were evaluated during summer season of 2015 at two locations [ICRISAT, Patancheru; and TNAU (Tamil Nadu Agricultural University), Coimbatore (11° N, 77° E, and 411.98 m above sea level, respectively)] for forage related morphological and quality traits. At ICRISAT, Patancheru, all entries were planted on alfisol in an alpha lattice design with three replications. Each entry was planted in 4 rows of 4 m length with rows spaced 60 cm apart and plants spaced at 10–12 cm from each other. At TNAU, Coimbatore, entries were planted in black soils with two replications. Each entry was planted in 3 rows, each of 2 m in length with rows spaced 45 cm apart. At both the locations, cultural practices were followed as described earlier under the head "phenotyping of hybrid parents"; and GFY, DFY, CP and IVOMD were again recorded as mentioned under the same heading.

2.6. Data Analysis

One hundred and thirty high biomass forage type hybrid parents (18 seed and 112 pollinator parents), including all the 116 which were phenotyped, were analyzed for SSRs. In the case of GBS-identified SNPs, eight parents were dropped due to missing information, so 122 (18 seed and 104 pollinator parents) hybrid parents were analyzed using 7870 (GBS-identified SNPs) markers. The allelic base pairs of all the markers were used to estimate the summary statistics, which include PIC, allelic richness as determined by total number of detected alleles and allele per locus, gene diversity and heterozygosity. Gene diversity, which is often referred to as expected heterozygosity, is defined as the probability that two randomly chosen alleles from the population are different. An unbiased estimator of gene diversity at the *l*th is $\hat{D}_l = \left(1 - \sum_{u=1}^{k} p_{lu}^2\right)/\left(1 - \frac{1+f}{n}\right)$, where p_{lu} is the frequency of the *u*th allele at the *l*th locus, n is the sample size, and f is the inbreeding depression, respectively. Occurrence of common, unique, rare and most frequent alleles were estimated using PowerMarker V3.25 software [45]. Common alleles, unique alleles and rare alleles were calculated as described by Li et al. [46] and Upadhaya et al. [47].

The genetic dissimilarities for each pair of forage type hybrid parents using simple matching coefficient matrix (SSRs) and Roger's distance matrix (GBS-identified SNPs) were calculated with the Power Marker V3.25 software. The simple match distance takes into account both the shared 0 s (absence of a band) and shared 1 s (presence of a band) as factors that contribute to similarity between two individuals. Simple matching distance was calculated as $d_{ij} = 1 - \frac{1}{L}\sum_{l=1}^{L}\frac{m_l}{n}$, where d_{ij} is the dissimilarity between units i and j, L is the number of loci, π is ploidy and m_l is the number of matching alleles for locus l, respectively. Rogers's distance was calculated as $D_R = \frac{1}{m} \sum_j^m \sqrt{\frac{1}{2} \sum_i^{a_j} \left(p_{ij} - q_{ij}\right)^2} \, p_{ij}$ and q_{ij} are the frequencies of ith allele at the jth locus in populations X and Y respectively, while a_j is the number of alleles at the jth locus, and m is the number of loci examined. Hybrid parents (seed and pollinator) were grouped by cluster analysis using the neighbor-joining method in Power Marker V3.25 software. The pairwise F_{st} method was used to infer the clusters in a neighbor-joining tree. Analysis of molecular variance (AMOVA) was performed to detect mutational differences between the loci in diverse group of populations by partitioning the variation within and between the hybrid parents [48]. AMOVA analysis provided the Wright's F statistic or F_{st} (Fixation index) based on genetic distance of the subgroups in both the marker systems. Association between hybrid parents were estimated with PCoA (Principal coordinate analysis) analysis using DARwin V6 software [49]. Mantel test was performed to investigate correlation between genetic distances estimated through SSRs and GBS-identified SNPs [50] based on 10,000 iterations using R software [51]. BLUP (Best Linear Unbiased Prediction) means were estimated for each of the 116 hybrid parents on the four traits evaluated and used for genetic relatedness based on forage traits using SAS v 9.4 [52].

Combined analysis of variance (ANOVA) for a hybrid estimation trial was estimated using SAS PROC MIXED [52]. The estimate of mid-parent and better parent heterosis for each trait was calculated from mean of two environments with the following formulae: Mid-parent heterosis (MPH, %) = $(F_1 - MP)/MP \times 100$; Better parent heterosis (BPH, %) = $(F_1 - BP)/BP \times 100$, where F_1 (First filial hybrid) is the hybrid mean, MP (Mid-parent) is the mean of both the parents, and BP (Better parent) is the superior parent over the other parent. The Pearson correlation coefficients (r) between GD and heterosis (MPH and BPH) were estimated using across environment trait means.

3. Results

3.1. Genetic Diversity Indicators Based on SSRs and GBS-Identified SNPs

Fifty two SSR markers detected a total of 551 alleles in 130 hybrid parents, with an average of 10.60 alleles per locus. The number of alleles per locus ranged from 2 (*Xpsmp2273*) to 31 (*Xpsmp2070*), with three to ten alleles at 29 SSR loci (Table S3). Markers *Xpsmp2068* (26), *Xpsmp2081.1* (30) and *Xpsmp2070* (31) had more than 20 alleles per locus. Furthermore, the average number of alleles per locus was higher for pollinator parents (9.81) than for seed parents (4.63) (Table 1). The polymorphism information content (PIC) values varied from 0.06 (*Xpsmp2267*) to 0.94 (*Xpsmp2070*) with a mean of 0.65. Pollinator parents had higher PIC values (0.63) than seed parents (0.52). The gene diversity varied from 0.06 (*Xpsmp2267*) to 0.94 (*Xpsmp2070*) with an average of 0.68. Moreover, pollinator parents showed higher gene diversity (0.66) than seed parents (0.55). The level of heterozygosity ranged from 0.0 to 0.12, with an average of 0.02. Of the 551 alleles, 129 were rare alleles (23.41%) ranging from 1 to 13, 341 were common alleles (61.89%) ranging from 1 to 23, and 81 were most frequent alleles (14.70%) ranging from 1 to 3. Genotype-specific (unique) alleles were detected in 3 seed parents and 14 pollinator parents (Table S4).

The average gene diversity was 0.48, which varied from 0.02 to 0.50 for GBS-identified SNP markers. The range of heterozygosity was 0.0 to 0.97, with an average of 0.15, while PIC varied from 0.02 to 0.38, with an average of 0.37 (Table 1).

Table 1. Average of allelic richness, major allele frequency, gene diversity, heterozygosity and polymorphism information content (PIC) of the 52 SSRs and 7870 GBS-identified SNPs in hybrid parents of forage pearl millet.

Groups	Allelic Richness [†]	Gene Diversity [†]	Heterozygosity [†]	PIC [†]
SSRs				
Seed parents	4.63 (1.00–13.00)	0.55 (0.00–0.90)	0.02 (0.00–0.14)	0.52 (0.00–0.89)
Pollinator parents	9.81 (2.00–29.00)	0.66 (0.05–0.90)	0.03 (0.00–0.17)	0.63 (0.05–0.93)
Seed and Pollinator parents	10.6 (2.00–31.00)	0.68 (0.06–0.94)	0.02 (0.00–0.12)	0.65 (0.06–0.94)
Standard error	0.90 (0.35–0.92)	0.03 (0.03–0.03)	0.00 (0.01–0.01)	0.03 (0.03–0.03)
GBS-Identified SNPs				
Seed parents	1.95 (1.00–2.00)	0.37 (0.00–0.50)	0.13 (0.00–1.00)	0.29 (0.00–0.38)
Pollinator parents	2.00 (1.00–2.00)	0.48 (0.00–0.50)	0.15 (0.00–0.96)	0.37 0.00–0.38)
Seed and Pollinator parents	2.00 (2.00–2.00)	0.48 (0.02–0.50)	0.15 (0.00–0.97)	0.37 (0.02–0.38)
Standard error	0.00 (0.00–0.00)	0.00 (0.00–0.00)	0.00 (0.00–0.00)	0.00 (0.00–0.00)

[†] Allelic richness, Genetic diversity, Heterozygosity and PIC value ranges for the seed and pollinator parents are given in the parentheses, respectively.

3.2. Clustering Pattern and Genetic Relatedness between Hybrid Parents Based on Markers

The neighbor-joining tree constructed based on simple matching (SSR based) distance matrix grouped all of the seed and pollinator parents into 2 clear-cut separate groups (statistical significance values provided in Table S5) (Figure 1a), each having majority of seed and pollinator hybrid parents. These broad groups were found to be partitioned further into three and four statistically distinct clusters of seed and pollinator parents, respectively (Tables S6 and S7). Twenty three pollinator parents were found grouped in clusters dominated by seed parents, of which only 7 were found clustered with seed parents, while 16 pollinators were grouped separately. Eight B-lines were found in B-I (44%), 6 in B-II (33%) and 4 in B-III (22%) (Figure 1b and Table S1). Cluster B-II had 67% (4 out of 6) of the lines derived directly or indirectly from ICMB 89111 (ICRISAT millet B-line) in their parentage. B-III cluster had 50% (2 out of 4) of the lines sharing 81B in parentage.

Pollinator parents found grouped into four clusters (Figure 1c and Table S1), of these 45 lines in R-I (40%), 24 in R-II (21%), 30 in R-III (27%) and 13 (12%) in R-IV. Twenty six out of 45 and 16 out of 24 pollinator parents in cluster I and II respectively, had ICMS 7704 (ICRISAT millet synthetic variety) in their parentage. ICMS 7704 is an open-pollinated variety developed from six inbred lines derived from Indian × African crosses selected at Tandojam in Pakistan. Seventeen out of 30 lines in R-III cluster had progenies derived from Medium composite 94 (MC 94) in their parentage; MC 94 is a medium maturity (75–85 days) composite developed at ICRISAT. R-IV cluster had 42% (5 out of 12) of the progenies sharing a common parent ICMS 8506.

Seed and pollinator parents followed almost the same clustering pattern based on both GBS-identified SNPs and SSRs (Figure 2a and Table S1) (Statistical significance values depicting distinctness of clusters provided in Tables S8–S10). Based on SNP based clustering, thirty-three pollinators were found grouped in clusters dominated by seed parents, but 30 of them were found to be grouped separately. SNP based clustering followed the same trend of pedigree linkages as found in case of SSRs; 33% (2 out of 6) and 67% (2 out of 3) of the lines in B-II and B-III respectively, had the ICMB 89111 parent (Figure 2b), while cluster R-I and R-II had 80% and 48% of the lines with ICMS 7704 respectively, and R-III had 57% (17 out of 30) of the lines with MC 94 in the parentage (Figure 2c).

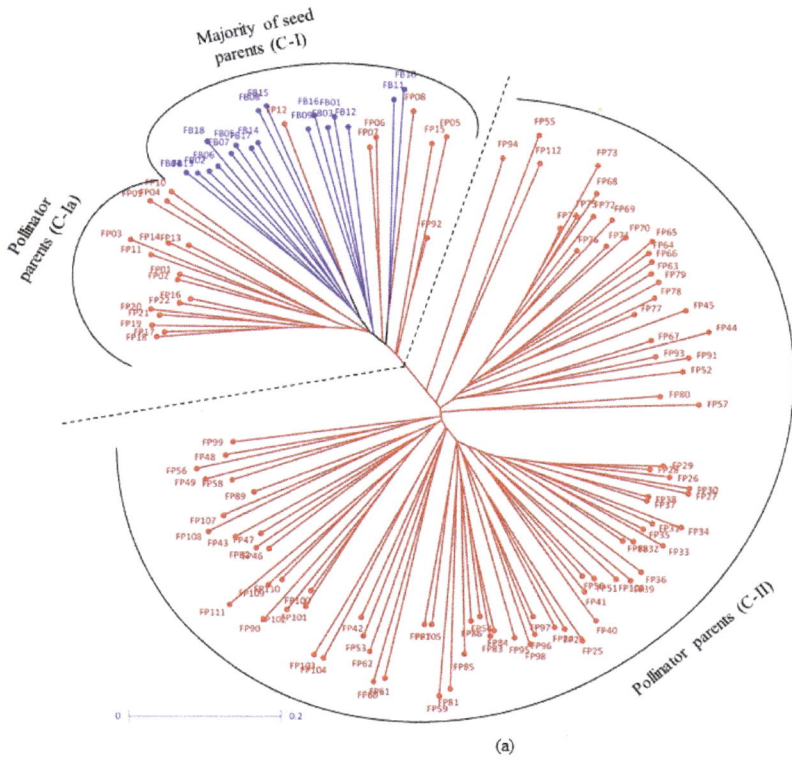

(a)

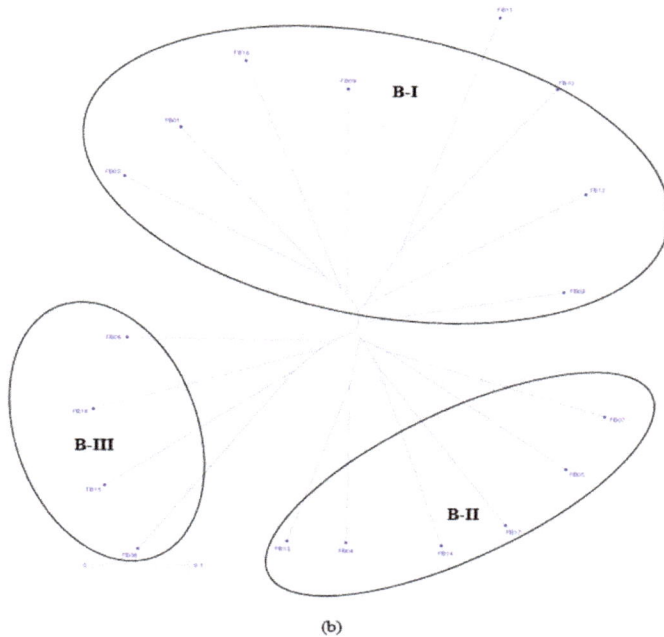

(b)

Figure 1. *Cont.*

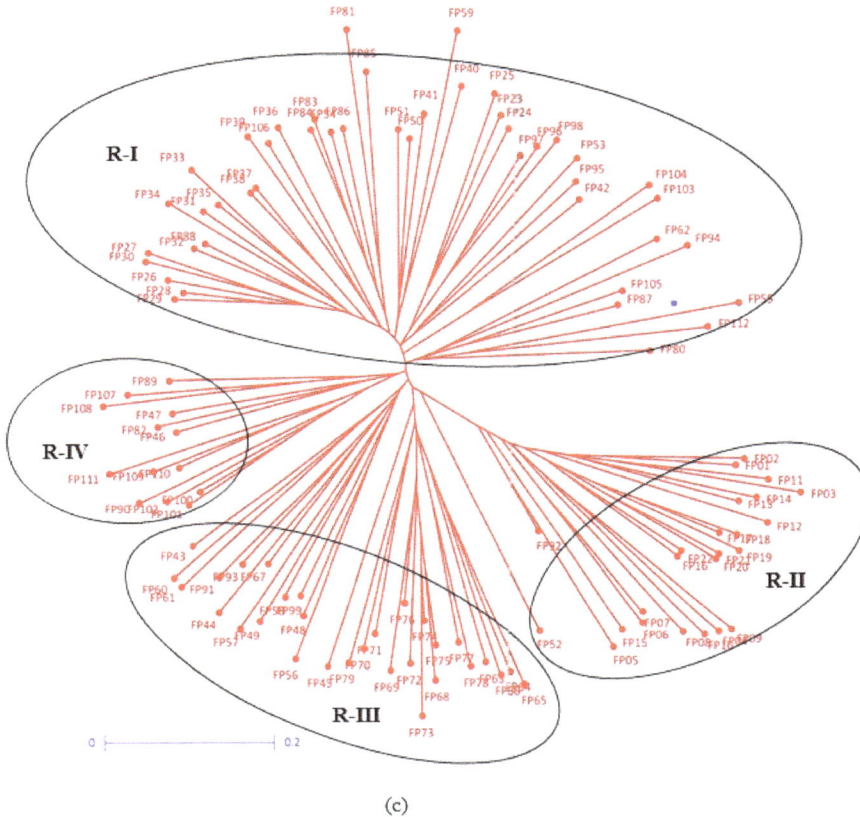

(c)

Figure 1. Neighbor joining tree based on 52 SSRs: (**a**) 130 seed and pollinator parents; (**b**) 18 seed parents; and (**c**) 112 pollinator parents.

AMOVA based on SSRs showed 83.62% of the variance within seed and pollinator parents and 16.38% between seed and pollinator parents ($p < 0.0001$). The F_{st} between seed and pollinator parents for individual markers varied from −0.0126 to 0.5381 and was significant for 41 of the 52 SSRs (Table S11). Some of the markers such as *Xpsmp*2201 (0.2567), *Xipes*0105 (0.2574), *Xipes*0004 (0.3227), *Xipes*0186 (0.3351), *Xpsmp*2214 (0.3445), *Xicmp*3048 (0.4848), *Xpsmp*2222 (0.5294) and *Xpsmp*2246 (0.5381) had the highest F_{st} values. Following the same trend, SNPs revealed 90.11% of the variance within seed and pollinator parents and 9.89% variance was observed between seed and pollinator parents (Table S12). PCoA based on SSR markers revealed first and second principal coordinates to account for 14.57% and 6.21% of the molecular variance, respectively (Table S13 and Figure S1a), while it was 33.63% and 7.74% based on the GBS-identified SNPs (Figure S1b). The Mantel test found significant association (r = 0.35, $p < 0.01$) between genetic distances (Simple matching distance for SSRs vs. Roger's distance for SNPs) estimated for hybrid parents.

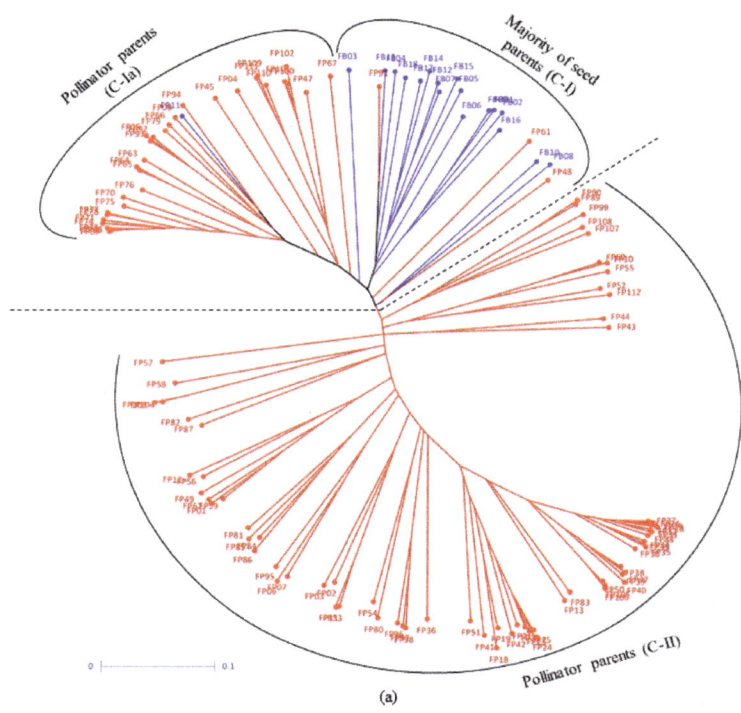

(a)

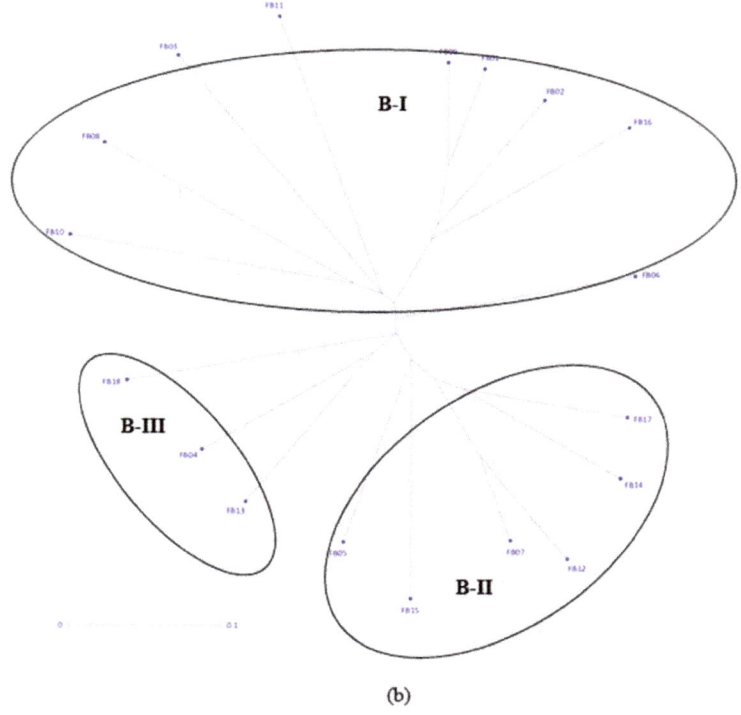

(b)

Figure 2. *Cont.*

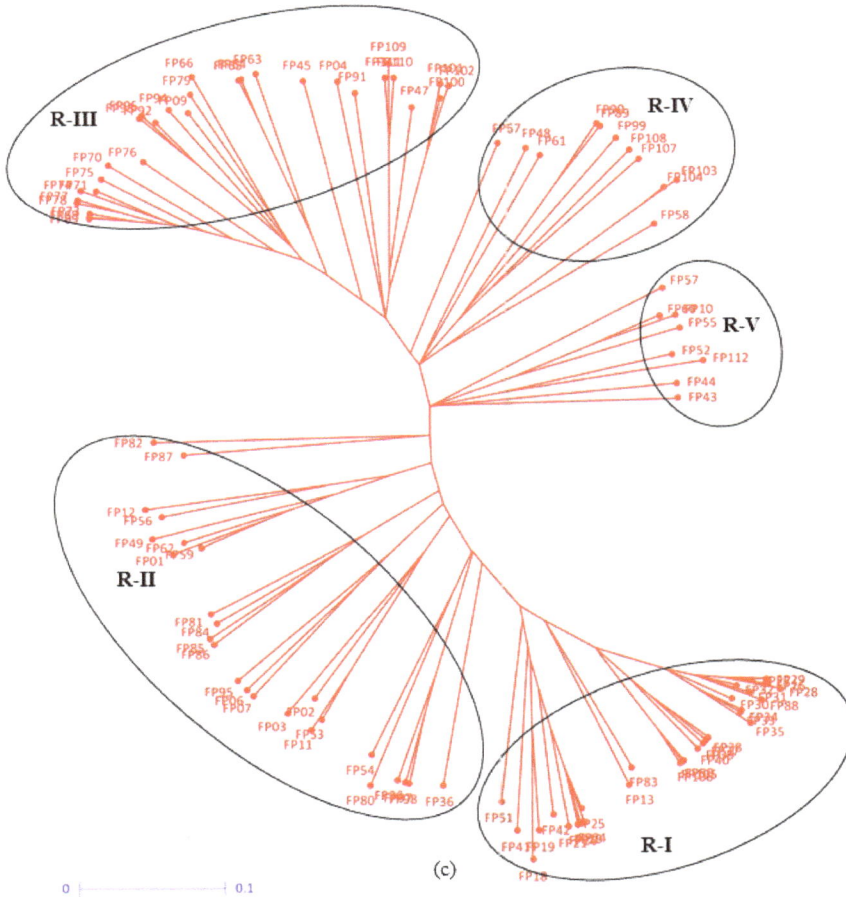

Figure 2. Neighbor joining tree based on 7870 GBS-identified SNPs: (**a**) 122 seed and pollinator parents; (**b**) 18 seed parents; and (**c**) 104 pollinator parents.

3.3. Variability for Forage Traits in Hybrid Parents

A wide range was observed in hybrid parents for GFY, DFY, CP and IVOMD at different cutting intervals (Table 2). GFY of hybrid parents at first cut ranged from 15 to 29 t ha^{-1} with an average of 22 t ha^{-1}, seed parents ranged from 16 to 23 t ha^{-1} with an average of 20 t ha^{-1} and pollinator parents ranged from 15 to 29 t ha^{-1} with an average of 22 t ha^{-1}. In addition, DFY varied from 3 to 6 t ha^{-1} with an average of 4 t ha^{-1} for hybrid parents, from 4 to 5 t ha^{-1} with an average of 4 t ha^{-1} for seed parents and from 3 to 6 t ha^{-1} with an average of 4 t ha^{-1} for pollinator parents. At the second cut, GFY of hybrid parents ranged from 12 to 42 t ha^{-1} with an average of 27 t ha^{-1}, seed parents ranged from 12 to 26 t ha^{-1} with an average of 20 t ha^{-1} and pollinator parents ranged from 16 to 42 t ha^{-1} with an average of 28 t ha^{-1}. Further, DFY varied from 5 to 9 t ha^{-1} with an average of 6 t ha^{-1} for both seed and pollinator parents, from 5 to 9 t ha^{-1} with an average of 6 t ha^{-1} for seed parents and from 5 to 8 t ha^{-1} with an average of 6 t ha^{-1} for pollinator parents. CP for hybrid parents, seed and pollinator parents ranged from 11% to 15%, 12% to 15% and 11% to 14% at the first cut, whereas it was 11% to 13%, 11% to 12% and 11% to 13% for second cut. IVOMD for hybrid parents ranged from 54% to 56%

at the first cut, while it ranged from 51% to 55% at the second cut. Seed parents varied from 54% to 55% at first cut and 51 to 54% at second cut, and pollinator parents varied from 54% to 56% and 51% to 55% at the first and second cuts, respectively.

Table 2. Mean and range of forage traits of 116 hybrid parents from 2 years mean data (summer 2015 and 2016).

Traits	Lines	First Cut		Second Cut	
		Mean	Range	Mean	Range
Green fodder yield (t ha^{-1})	Seed parents	20.32	16.30–22.89	19.79	12.35–26.21
	Pollinator parents	22.21	15.18–29.16	28.03	15.61–42.42
Dry fodder yield (t ha^{-1})	Seed parents	4.25	3.63–5.19	6.21	5.23–8.99
	Pollinator parents	4.42	3.34–5.91	6.31	5.06–8.09
Crude protein (%)	Seed parents	12.92	11.94–14.61	11.86	11.40–12.44
	Pollinator parents	12.55	11.49–13.52	11.71	11.04–12.51
In vitro organic matter digestibility (%)	Seed parents	54.66	53.93–55.40	52.67	51.16–54.26
	Pollinator parents	54.81	53.78–55.63	52.58	51.01–55.04

3.4. Trait Association with SSRs Based Clusters

Forage quantity and quality traits associated with SSRs based clusters revealed that B-III cluster had the highest mean values for GFY and DFY at the first cut and the highest GFY at the second cut, while DFY at second cut was found highest in cluster B-I (Table 3). Meanwhile for pollinator parent clusters, R-I had highest mean for GFY at first cut and for GFY and DFY at second cut. Cluster B-II had highest CP and IVOMD in first cut; while B-III had highest CP in second cut. Cluster R-II and R-III had the highest CP in the first and second cut, respectively. For seed parental clusters, B-I had the highest IVOMD in the second cut. Cluster R-II and R-III had the highest IVOMD in the first cut and R-IV in the second cut.

3.5. ANOVA for Heterosis Estimation Trial

ANOVA for forage yield and quality (Table 4) traits showed large and highly significant variance due to locations, indicating that the materials were evaluated under diverse environments. Large and highly significant variance observed in either of the parents (seed or pollinator parent) and in hybrids for almost all the traits indicated wide genetic differences among parental lines as well as among hybrids. High significant mean squares due to "hybrids vs. parents" (except IVOMD in second cut) indicated significant heterosis in hybrids for most of the forage traits. Forage yield performance of both parental lines and hybrids (except for DFY in first cut) was significantly modified by environments. The significant and relatively large percentage of the total variation attributable to the G × E (Genotype × Environment) suggested that hybrids responded differentially to environments for forage traits.

Table 3. Mean values for forage traits in different SSR-based clusters on the basis of 2 year data (Summer season at ICRISAT, 2015 and 2016).

Lines	Cluster No.	First Cut				Second Cut			
		GFY (t ha⁻¹) [†]	DFY (t ha⁻¹) [†]	CP (%) [†]	IVOMD (%) [†]	DFY (t ha⁻¹) [†]	GFY (t ha⁻¹) [†]	CP (%) [†]	IVOMD (%) [†]
Seed parents	B-I	20.14 (17.13, 22.39)	4.21 (3.63, 4.46)	12.75 (11.94, 13.47)	54.69 (54.04, 55.40)	**6.31** (5.23, 8.99)	18.72 (12.35, 22.02)	11.82 (11.56, 12.13)	**52.90** (51.80, 54.26)
	B-II	19.67 (16.30, 22.89)	4.10 (3.70, 4.54)	**13.07** (12.22, 14.61)	**54.79** (54.47, 55.03)	6.16 (5.98, 6.43)	20.53 (14.57, 26.21)	11.82 (11.40, 12.44)	52.68 (52.01, 53.70)
	B-III	**21.63** (20.60, 22.33)	**4.54** (3.87, 5.19)	13.02 (12.57, 13.49)	54.43 (53.93, 55.06)	6.09 (6.02, 6.25)	**20.84** (17.30, 23.81)	**12.01** (11.81, 12.13)	52.18 (51.16, 53.30)
	R-I	**22.81** (18.44, 27.40)	4.46 (3.64, 5.61)	12.40 (11.49, 13.18)	54.79 (53.78, 55.56)	**6.44** (5.47, 7.69)	**29.73** (15.61, 37.30)	11.67 (11.04, 12.36)	52.41 (51.01, 53.83)
Pollinator parents	R-II	21.22 (15.18, 25.49)	4.29 (3.41, 4.80)	**12.75** (12.23, 13.52)	**54.85** (54.23, 55.47)	6.21 (5.06, 6.94)	26.36 (16.35, 34.75)	11.69 (11.21, 12.04)	52.60 (51.33, 53.69)
	R-III	22.24 (17.48, 29.16)	4.42 (3.69, 5.91)	12.64 (11.77, 13.42)	**54.85** (54.03, 55.63)	6.28 (5.31, 8.09)	27.99 (19.56, 42.42)	**11.78** (11.13, 12.51)	**52.77** (51.60, 55.04)
	R-IV	21.97 (19.89, 24.07)	4.53 (3.34, 5.51)	12.52 (11.55, 13.12)	54.68 (54.21, 55.15)	6.06 (5.17, 7.21)	23.29 (18.83, 28.89)	11.72 (11.35, 12.38)	**52.91** (51.85, 54.27)

[†] Average values; minimum and maximum values in parenthesis, respectively, of the trait within the cluster, Values are in bold indicates higher mean value for the trait, GFY-Green forage yield, DFY-Dry forage yield, CP-Crude protein, IVOMD-*In vitro* organic matter digestibility.

Table 4. Analysis of variance for combining ability for forage related morphological and biochemical traits at first and second cut, evaluated in hybrids and parental trial at two locations in summer season of 2015.

Source of Variation	DF [†]FC	DF [‡]SC	GFY FC	GFY SC	DFY FC	DFY SC	CP FC	CP SC	IVOMD FC	IVOMD SC
Locations	1	1	5.71 *	309.26 ***	277.36 ***	97.57 ***	116.88 ***	25.07 ***	562.89 ***	987.69 ***
Replication (Locations)	3	3	4.43 *	2.63	6.85 ***	1.27	3.47 **	5.57 **	0.29	6.00 **
Treatments	97	58	6.18 ***	11.08 ***	2.58 ***	3.44 ***	4.73 ***	2.56 ***	1.55 *	3.59 ***
Parents	17	13	3.37 ***	5.98 ***	0.11	1.65	6.98 ***	3.00 ***	1.64	6.17 ***
(i) Females	9	8	1.12	2.86	0.31	0.68	8.08 ***	3.18 ***	2.21 *	3.71 ***
(ii) Males	7	4	21.19 ***	9.15 ***	1.22	3.02 *	4.42 ***	2.35	1.12	5.15 ***
(iii) Females vs. Males	1	1	264.05 ***	21.50 ***	141.81 ***	4.37 *	12.99 ***	3.35	0.22	33.7 ***
Hybrids	79	44	11.13 ***	8.08 ***	1.42	2.57 ***	2.65 ***	1.65 *	1.44 *	2.88 ***
(i) Lines	9	8	6.96	2.16 *	0.96	1.65	3.63 ***	1.48	3.13 ***	4.84 ***
(ii) Testers	7	4	0.97	21.82 ***	1.05	11.53 ***	6.92 ***	6.04 ***	2.06 *	4.94 ***
(iii) Lines × Testers	63	32	3.49 ***	7.04 ***	0.23	1.64 *	2.15 ***	1.04	1.24	2.17 ***
Hyb vs. Par	1	1	1.46 *	190.82 ***	1.83 ***	61.11 ***	108.76 ***	29.97 ***	7.70 **	1.77
Location × Treatments	97	58	2.76 ***	6.12 ***	1.09	1.84 ***	2.48 ***	1.68 *	1.76 ***	2.95 ***
Location × Parents	17	13	0.87	4.08 ***	0.21	0.96	2.47 ***	1.29	0.75	4.02 ***
(i) Location × Females	9	8	0.94	3.54 ***	0.08	0.92	3.29 ***	0.89	0.41	4.59 ***
(ii) Location × Males	7	4	0.55	6.03 ***	0.4	1.28	1.67	1.65	1.17	0.82
(iii) Location × (Lines vs. Testers)	1	1	2.37	1	0.03	0.01	1.07	3.51	0.94	13.35 ***
Location × Hybrids	79	44	1.22	5.71 ***	1.03	2.01 ***	2.32 ***	1.02	1.33	2.42 ***
(i) Location × Lines	9	8	1.01	1.41	0.74	0.96	2.53 **	0.96	2.58 **	2.39 *
(ii) Location × Testers	7	4	0.91	10.24 ***	0.31	7.35 ***	2.89 **	3.80 **	0.81	0.75
(iii) Location × (Lines × Testers)	63	32	1.29	5.82 ***	1.14	1.53 *	2.27 ***	0.64	1.19	2.59 ***
Location × (Hybrids vs. Parents)	1	1	28.05 ***	45.02 ***	79.69 ***	6.21 **	12.76 ***	34.65 ***	50.91 ***	16.97 ***

DF-Degree of freedom; [†] First cut; [‡] Second cut, GFY-Green forage yield (t ha^{-1}), DFY-Dry forage yield (t ha^{-1}), CP-Crude protein (%), IVOMD-*In vitro* organic matter digestibility (%). *, **, *** Significant at 0.05, 0.01 and 0.001 level, respectively.

3.6. Magnitude of Heterosis and Its Association with GD

The extent of MPH and BPH for forage quantity and quality traits over two cuts are presented in Table 5. The MPH for DFY ranged from 33% to 407%, with an average of 189%, while BPH ranged from −13% to 344%, with an average of 154% at the first cut. Similarly, MPH for IVOMD ranged from −7% to 7%, with an average of 2%, while BPH ranged from −8% to 7% at first cut. The correlation between SSR based GD and heterosis (MPH and BPH) for forage related morphological and quality traits are given in Table 6. Low positive correlation was observed between GD and MPH ($r = 0.34$, $p < 0.05$), and with BPH ($r = 0.41$, $p < 0.05$) for GFY at the second cut; GD and BPH ($r = 0.33$, $p < 0.05$) for DFY at the second cut and; GD and MPH ($r = 0.26$, $p < 0.05$) for CP at the first cut. No significant correlation was found between GD and heterosis (both MPH and BPH) for IVOMD for both the cuts. Similarly, for SNPs, significant positive relationships were found between GD and MPH ($r = 0.32$, $p < 0.01$) and with BPH ($r = 0.28$, $p < 0.05$) for DFY at the first cut.

Table 5. Summary of MPH and BPH for forage linked traits.

Traits	Cutting Intervals	Mid-Parent Heterosis (%)			Better-Parent Heterosis (%)		
		Minimum	Maximum	Average	Minimum	Maximum	Average
GFY	First cut	20.6	115.7	58.2	−8.2	74.4	32.6
	Second cut	−22.2	378.1	103.8	−37.0	301.9	51.5
DFY	First cut	33.1	406.5	189.1	−13.1	344.3	154.2
	Second cut	−15.5	290.0	93.2	−29.7	248.2	50.7
CP	First cut	−23.6	9.8	−9.5	−27.8	8.8	−14.4
	Second cut	−24.1	7.2	−7.8	−30.7	4.5	−12.8
IVOMD	First cut	−6.8	7.1	1.3	−8.2	6.7	0.1
	Second cut	−7.5	5.9	0.0	−10.8	3.4	−2.9

GFY-Green forage yield (t ha^{-1}), DFY-Dry forage yield (t ha^{-1}), CP-Crude protein (%), IVOMD-*In vitro* organic matter digestibility (%).

Table 6. Correlation between genetic distance (GD) between parents measured using SSRs and SNPs and heterosis (Mid-parent heterosis (MPH) and Better-parent heterosis (BPH)) for forage yield and quality traits in pearl millet.

Traits	Cutting Intervals	Correlation Coefficient between GD and MPH		Correlation Coefficient between GD and BPH	
		SSRs	SNPs	SSRs	SNPs
Green forage yield (GFY, t ha^{-1})	First cut	0.11	−0.02	0.12	−0.15
	Second cut	0.34 *	0.15	0.41 *	0.13
Dry forage yield (DFY, t ha^{-1})	First cut	−0.10	0.32 **	−0.14	0.28 *
	Second cut	0.25	0.19	0.33 *	0.19
Crude protein (CP, %)	First cut	0.26 *	0.04	0.20	−0.12
	Second cut	−0.05	−0.22	−0.02	−0.19
In vitro organic matter digestibility (IVOMD, %)	First cut	0.11	0.00	0.18	−0.01
	Second cut	−0.18	0.06	−0.17	0.11

*, ** Significant at 0.05 and 0.01 level, respectively.

4. Discussion

The average number of alleles per locus (10.60) detected in this study was higher than earlier reports in pearl millet, 6.26 alleles per locus in 72 inbred lines [53] and 2.76 alleles per locus in 42 inbred lines [54]. These differences might be due to the lesser number of parental lines and limited number of SSRs (25 to 34) in those earlier conducted studies on grain type hybrid parents, whereas the current study had a higher number of diverse forage type hybrid parents (130) and also had a higher number of microsatellite markers (52 SSRs). However, it was still lower in comparison to other studies, like of

Stich et al. [13] (16.4 alleles per locus in 145 inbred lines derived from landraces and open-pollinated varieties of West and Central Africa) and Gupta et al. [15] (12.68 alleles per locus in 379 inbred lines), which might be due to the involvement of a comparatively lesser number of hybrid parents (130) in this study. The present study detected a higher average number of alleles per locus, PIC and gene diversity in pollinator parents than seed parents, indicating that pollinator parents were genetically more diverse than seed parents. This might be due to the broader genetic base of germplasm used in the development of pollinator parents and also due to differences in the sample size (112 pollinator parents vs. 18 seed parents). These findings are in consonance with earlier investigations in pearl millet [14,15], which reported high diversity in pollinator parents than in seed parents.

Majority of seed and pollinator parents delineated into two separate groups based on SSRs. Similar kinds of clear cut separate seed and pollinator grouping patterns were reported earlier in pearl millet [14,15] and in rice [55]. Distributions of hybrid parents under study into different marker-based groups indicated presence of significant genetic diversity in the breeding materials. Also, most of parents found in the common cluster had the involvement of some common parent in the pedigree. Such genetic relatedness between hybrid parents as found in SSR based clusters was also reported earlier by Nepolean et al. [14] and Gupta et al. [15] in pearl millet hybrid parents bred for grain-type traits. The clustering pattern of hybrid parents and trend of involvement of common parents in parentage of lines in different clusters were almost similar for both type of marker systems, which might be due to significant positive correlation ($r = 0.35$, $p < 0.01$) found between parental genetic distance assessed using SSRs and GBS-identified SNPs markers. However, seed parental cluster B-III had 81B and ICMB 89111 in their parentage for SSRs and GBS-identified SNPs, respectively. Pollinator parents were grouped into four (SSRs) and five (SNPs) clusters, respectively. Pollinator parental group R-IV (SSR) had the lines sharing ICMS 8506 in parentage while SNP based R-IV had HHVBC tall in their parentage. Hence, our study suggested that any of the two marker systems, SSRs or SNPs, can be effectively used for diversity investigations for forage type pearl millet breeding materials in future. High correlation between genetic distances estimated by SSRs and SNPs was also reported earlier in sunflower [56], while no such correlation was found in maize [57,58].

The study revealed significant genetic variability for forage quantity and quality traits in hybrid parents under investigation. Several other studies have also reported a wide range of green forage yield in pearl millet [12,59–62]. Earlier, Rai et al. [63] also reported dry forage yield in hybrid parents in a range of 3 to 4 t ha^{-1} at the first cut. Stover CP varied from 11% to 15% and 11% to 13% at the first and second cut, respectively, which was more than the minimum (about 7%) required by rumen microbes [64]. The observed genotypic variations in crude protein in high biomass hybrid parents in this study can also help to enhance the capacity of feed intake, as suggested earlier by Van Soest [64].

The mean values observed for forage quality traits (CP and IVOMD) at the first and second cuts were higher than observed in previous studies in pearl millet [11,12,44,63,65,66] and also in several other crops, like in maize [67,68] and in sorghum [69]. Variability in IVOMD assumes high significance as a one-percent unit increase in digestibility in stover sorghum and pearl millet could result in increases in milk, meat and draught power outputs in a range from 6% to 8% [70]. Market studies on sorghum fodder also indicated that the pricing of fodder is affected by its quality traits, especially for IVOMD [71]. In hybrid parents involved in this study, stover IVOMD varied from 54% to 56% units at first cut, and from 51% to 55% in second cut. Such a wide range of IVOMD in forage type hybrid parental lines can be utilized in the breeding program for the development of quality cultivars with high digestibility.

Mean values of forage traits in SSR based clusters revealed significant variation across clusters. The study identified trait-specific clusters; such associations between marker-based clusters and forage traits might be due to a close association of markers under investigation with the specific forage trait. This study has not done association mapping between markers and forage traits, so the associations observed in this study might be the result of stratification rather than true association, but diverse hybrid parents from these trait-specific clusters can still provide inputs to breeders for utilization in forage trait-specific targeted breeding program.

Hybrids showed an average heterosis of 154% and 51% over better parent for DFY in the first and second cut, respectively, indicating that higher heterosis can be realized for forage traits in pearl millet. These results pointed that high yielding multi-cut forage type hybrids with better forage quality can be developed in pearl millet for enhancing livestock productivity [63,65]. However, a high G×E for forage traits in hybrids indicated that there is a need to breed agro-ecological specific forage type hybrids to have stable forage yields. Significant but low positive correlation was found between GD and heterosis for most of the forage quality traits in either first or second cut, suggesting that SSR-based GD can be used for predicting heterosis for GFY, DFY and CP in pearl millet. Again, SNPs based GD also found a significant association with heterosis for DFY. In contrast to our results, Gupta et al. [17] found no significant positive correlation with BPH for DFY in either of morphological or molecular distance in pearl millet. Low correlation between GD and heterosis might be due to a lack of linkage between genes for traits under investigation, unequal genome coverage, and random marker distribution in the genome [72,73].

5. Conclusions

In conclusion, this molecular marker based genetic diversity study indicated that available forage-type pearl millet hybrid parents exist as two separate gene pools, one each for seed and pollinator parents. Significant genetic diversity exists among parents for forage quantity and quality traits. The study also attempted to find associations between marker-based clustering patterns and forage quantity and quality traits, to enable breeders to generate crosses among diverse forage type hybrid parents belonging to distinct clusters and develop hybrids and parental lines with higher potential. High heterosis observed for forage quality traits indicated possibilities to develop exclusive forage type hybrids with more and better forage quality. Low positive but significant correlation between markers based genetic distance and heterosis for most of the forage traits indicated that forage traits can be predicted to some extent in pearl millet.

Supplementary Materials: The following are available online at http://www.mdpi.com/2077-0472/9/5/97/s1, Figure S1: Principal coordinate analysis of seed and pollinator parents based on (a) SSRs and (b) GBS-identified SNPs, Table S1: Pedigree of 130 forage type hybrid parents, cluster numbers based on SSR and SNP markers, and selected parents for hybrid development used in the current study, Table S2: List of markers (Forward/Reverse primer), linkage group and repeat motifs of 52 SSRs used in the study, Table S3: Allelic richness, Polymorphism Information Content (PIC), gene diversity and heterozygosity of the 52 SSRs in 18 seed parents and 112 restorer parents of forage type pearl millet, Table S4: Genotype-specific alleles present in 3 seed parents and 14 pollinator parents of forage pearl millet, Table S5: Pairwise Fst values for 130 hybrid (18 seed and 112 pollinator) parents grouped based on SSRs, Table S6: Pairwise Fst values for 18 seed parents grouped based on SSRs, Table S7: Pairwise Fst values for 112 pollinator parents grouped based on SSRs, Table S8: Pairwise Fst values for 122 hybrid (18 seed and 104 pollinator) parents grouped based on GBS-identified SNPs, Table S9: Pairwise Fst values for 18 seed parents grouped based on GBS-identified SNPs, Table S10: Pairwise Fst values for 104 pollinator parents grouped based on GBS-identified SNPs, Table S11: Significant test values for Fst from Analysis of molecular variance (AMOVA) for 52 SSRs, Table S12: Analysis of molecular variance of SSRs and GBS-identified SNPs, Table S13: Principal coordinate analysis estimated percentage of variation explained by two axes using SSRs and GBS-identified SNPs.

Author Contributions: S.K.G., M.M., S.P. and M.B. designed the experiments. A.R. and R.R.D. helped in data analysis. M.B. involved in phenotyping for forage quality traits by NIRS. P.G.G.P. and S.K.G. wrote the paper. M.M. and S.P. reviewed the manuscript. All authors read and approved the final manuscript.

Funding: Financial support of the project was provided by the ICRISAT-Pearl Millet Hybrid Parents Research Consortium (PMHPRC) and conducted under CGIAR, Research Program on Dryland Cereals to the first author to carry out this research work as a part of doctoral program.

Acknowledgments: This research work was carried out as part of a Ph.D. thesis of Govintharaj, submitted to Tamil Nadu Agricultural University, Coimbatore, Tamil Nadu. It was financially supported by the ICRISAT-Pearl Millet Hybrid Parents Research Consortium (PMHPRC) and conducted under CGIAR, Research Program on Dryland Cereals. We also acknowledge Mahender Thudi from the ICRISAT-CEG for his assistance in lab work. We also thank K.V.S. Prasad and Ramakrishna Reddy from ICRISAT-ILRI for sample preparation and NIRS analysis.

Conflicts of Interest: The authors declare no conflict of interest.

Abbreviations

SSR	Simple sequence repeats
GBS	Genotyping by sequencing
SNP	Single nucleotide polymorphism
GFY	Green forage yield
CP	Crude protein
IVOMD	*In vitro* organic matter digestibility
NIRS	Near infra-red reflectance spectroscopy
GD	Genetic distance
DFY	Dry forage yield
ME	Metabolizable energy
ICRISAT	International Crops Research Institute for the Semi-Arid Tropics
FB	Forage B line
FP	Forage pollinator
CMS	Cytoplasmic male sterility
PCR	Polymerase chain reaction
UNEAK	Universal Network Enabled Analysis Kit
MAF	Minor allele frequency
TNAU	Tamil Nadu Agricultural University
AMOVA	Analysis of molecular variance
Fst	Fixation index
PCoA	Principal coordinate analysis
BLUP	Best Linear Unbiased Prediction
ANOVA	Analysis of variance
MPH	Mid-parent heterosis
BPH	Better parent heterosis
MP	Mid parent
BP	Better parent
F1	First filial hybrid
PIC	Polymorphism information content
ICMB	ICRISAT millet B-line
ICMS	ICRISAT millet synthetic composite variety
MC	Medium composite
G × E	Genotype × Environment

References

1. Burton, G. History of hybrid development in pearl millet in Tifton. In Proceedings of the First Grain Pearl Millet Symposium, Tifton, GA, USA, 17–18 January 1995; pp. 5–8.
2. Davis, A.J.; Dale, N.M.; Ferreira, F.J. Pearl millet as an alternative feed ingredient in broiler diets. *J. Appl. Poultry Res.* **2003**, *12*, 137–144. [CrossRef]
3. Hanna, W. Improvement of millets: Emerging trends. In *Proceedings of the 2nd International Crop Science Congress, New Delhi, India*; Chopra, V.L., Singh, R.B., Varma, A., Eds.; Oxford & IBH Publishing Co. Pvt. Ltd.: New Delhi, India, 1996; pp. 139–146.
4. Reddy, A.A.; Malik, D.; Singh, I.P.; Ardeshna, N.J.; Kundu, K.K.; Rao, P.; Gupta, S.K.; Sharma, R.; Gajanan, G.N. Demand and supply for pearl millet Grain and fodder by 2020 in Western India. *Agric. Situat. India* **2012**, *68*, 635–646.
5. Amarender Reddy, A.; Yadav, O.; Dharm Pal Malik, S.I.; Ardeshna, N.; Kundu, K.; Gupta, S.K.; Sharma, R.; Sawarganokar, G.L.; Shyam, M.; Reddy, K.S. *Utilization Pattern, Demand and Supply of Pearl Millet Grain and Fodder in Western India*; Working Paper Series No. 37; ICRISAT: Patancheru, India, 2013.
6. De Assis, R.L.; de Freitas, R.S.; Mason, S.C. Pearl millet production practices in Brazil: A review. *Exp. Agric.* **2018**, *54*, 699–718. [CrossRef]

7. Dias-Martins, A.M.; Pessanha, K.L.; Pacheco, S.; Rodrigues, J.A.S.; Carvalho, C.W.P. Potential use of pearl millet (*Pennisetum glaucum* (L.) R. Br.) in Brazil: Food security, processing, health benefits and nutritional products. *Food Res. Int.* **2018**, *109*, 175–186. [CrossRef]

8. IGFRI (Indian Grassland and Fodder Research Institute). Vision 2050. 2013. Available online: http://www.igfri.res.in/ (accessed on 28 August 2018).

9. Rai, K.; Yadav, O.; Gupta, S.K.; Mahala, R. Emerging research priorities in pearl millet. *J. SAT Agric. Res.* **2012**, *10*, 1–5.

10. Rai, K.N.; Reddy, B.V.S.; Dakheel, A.J. Forage potential of sorghum and pearl millet adapted to dry lands. In *Proceedings of the Seventh International Conference on the Development of Drylands, Tehran, Iran, 14–17 September 2003*; Beltagy, K.B., Saxena, Eds.; Sustainable Development and Management of Drylands in the Twenty-First Century; International Centre for Agricultural Research in the Dry Areas: Aleppo, Syria, 2005; pp. 243–251.

11. Bidinger, F.; Blummel, M.; Hash, C.; Choudhary, S. Genetic enhancement for superior food-feed traits in a pearl millet (*Pennisetum glaucum* (L.) R. Br.) variety by recurrent selection. *Anim. Nutr. Feed Technol.* **2010**, *10*, 61–68.

12. Gupta, S.; Ghouse, S.; Atkari, D.; Blümmel, M. Pearl millet with higher stover yield and better forage quality: Identification of new germplasm and cultivars. In Proceedings of the 3rd Conference of Cereal Biotechnology and Breeding/CBB3, Berlin, Germany, 2–4 November 2015.

13. Stich, B.; Haussmann, B.I.; Pasam, R.; Bhosale, S.; Hash, C.T.; Melchinger, A.E.; Parzies, H.K. Patterns of molecular and phenotypic diversity in pearl millet [*Pennisetum glaucum* (L.) R. Br.] from West and Central Africa and their relation to geographical and environmental parameters. *BMC Plant Biol.* **2010**, *10*, 216. [CrossRef]

14. Nepolean, T.; Gupta, S.; Dwivedi, S.; Bhattacharjee, R.; Rai, K.; Hash, C. Genetic diversity in maintainer and restorer lines of pearl millet. *Crop Sci.* **2012**, *52*, 2555–2563. [CrossRef]

15. Gupta, S.K.; Nepolean, T.; Sankar, S.M.; Rathore, A.; Das, R.R.; Rai, K.N.; Hash, C.T. Patterns of molecular diversity in current and previously developed hybrid parents of pearl millet [*Pennisetum glaucum* (L.) R. Br.]. *Am. J. Plant Sci.* **2015**, *6*, 1697–1712. [CrossRef]

16. Hu, Z.; Mbacké, B.; Perumal, R.; Guèye, M.C.; Sy, O.; Bouchet, S.; Vara Prasad, P.V.; Morris, G.P. Population genomics of pearl millet (*Pennisetum glaucum* (L.) R. Br.): Comparative analysis of global accessions and Senegalese landraces. *BMC Genom.* **2015**, *16*, 1048. [CrossRef] [PubMed]

17. Gupta, S.K.; Nepolean, T.; Shaikh, C.G.; Rai, K.; Hash, C.T.; Das, R.R.; Rathore, A. Phenotypic and molecular diversity-based prediction of heterosis in pearl millet (*Pennisetum glaucum* L. (R.) Br.). *Crop J.* **2018**, *6*, 271–281. [CrossRef]

18. Ramya, A.R.; Ahamed, M.; Satyavathi, C.T.; Rathore, A.; Katiyar, P.; Raj, A.G.B.; Kumar, S.; Gupta, R.; Mahendrakar, M.D.; Yadav, R.S.; et al. Towards defining heterotic gene pools in pearl millet [*Pennisetum glaucum* (L.) R. Br.]. *Front. Plant Sci.* **2018**, *8*, 1934. [CrossRef]

19. Sattler, F.; Sanogo, M.; Kassari, I.; Angarawai, I.; Gwadi, K.; Dodo, H.; Haussmann, B.I.G. Characterization of West and Central African accessions from a pearl millet reference collection for agro-morphological traits and *Striga* resistance. *Plant Genet. Resour.-C* **2018**, *16*, 260–272. [CrossRef]

20. Singh, S.; Gupta, S.; Thudi, M.; Das, R.R.; Vemula, A.; Garg, V.; Varshney, R.K.; Rathore, A.; Pahuja, S.K.; Yadav, D.V. Genetic Diversity Patterns and Heterosis Prediction Based on SSRs and SNPs in Hybrid Parents of Pearl Millet. *Crop Sci.* **2018**, *58*, 2379–2390. [CrossRef]

21. Chowdari, K.; Venkatachalam, S.; Davierwala, A.; Gupta, V.; Ranjekar, P.; Govila, O. Hybrid performance and genetic distance as revealed by the (GATA) 4 microsatellite and RAPD markers in pearl millet. *Theor. Appl. Genet.* **1998**, *97*, 163–169. [CrossRef]

22. Betran, F.; Beck, D.; Bänziger, M.; Edmeades, G. Genetic analysis of inbred and hybrid grain yield under stress and nonstress environments in tropical maize. *Crop Sci.* **2003**, *43*, 807–817. [CrossRef]

23. Wegary, D.; Vivek, B.; Labuschagne, M. Association of parental genetic distance with heterosis and specific combining ability in quality protein maize. *Euphytica* **2013**, *191*, 205–216. [CrossRef]

24. Cai, J.; Lan, W. Using of AFLP marker to predict the hybrid yield and yield heterosis in rice. *Chin. Agric. Sci. Bull.* **2005**, *21*, 39.

25. Zhao, Q.Y.; Zhu, Z.; Zhang, Y.D.; Zhao, L.; Chen, T.; Zhang, Q.F.; Wang, C.L. Analysis on correlation between heterosis and genetic distance based on simple sequence repeat markers in japonica rice. *Chin. J. Rice Sci.* **2009**, *23*, 141–147.

26. Darvishzadeh, R. Phenotypic and molecular marker distance as a tool for prediction of heterosis and F_1 performance in sunflower (*Helianthus annuus'* L.) under well-watered and water-stressed conditions. *Aust. J. Crop Sci.* **2012**, *6*, 732.

27. Dreisigacker, S.; Melchinger, A.; Zhang, P.; Ammar, K.; Flachenecker, C.; Hoisington, D.; Warburton, M.L. Hybrid performance and heterosis in spring bread wheat, and their relations to SSR-based genetic distances and coefficients of parentage. *Euphytica* **2005**, *144*, 51–59. [CrossRef]

28. Ndhlela, T.; Herselman, L.; Semagn, K.; Magorokosho, C.; Mutimaamba, C.; Labuschagne, M.T. Relationships between heterosis, genetic distances and specific combining ability among CIMMYT and Zimbabwe developed maize inbred lines under stress and optimal conditions. *Euphytica* **2015**, *204*, 635–647. [CrossRef]

29. Xie, F.; He, Z.; Esguerra, M.Q.; Qiu, F.; Ramanathan, V. Determination of heterotic groups for tropical Indica hybrid rice germplasm. *Theor. Appl. Genet.* **2014**, *127*, 407–417. [CrossRef] [PubMed]

30. Wang, K.; Qiu, F.; Larazo, W.; dela Paz, M.A.; Xie, F. Heterotic groups of tropical *indica* rice germplasm. *Theor. Appl. Genet.* **2015**, *128*, 421–430. [CrossRef] [PubMed]

31. Gvozdenović, S.; Saftić-Panković, D.; Jocić, S.; Radić, V. Correlation between heterosis and genetic distance based on SSR markers in sunflower (*Helianthus annus* L.). *J. Agric. Sci.* **2009**, *54*, 1–10.

32. Burton, G.W. Pearl millets Tift 23DA and Tift 23DB released. *Georgia Agric. Res.* **1967**, *9*, 6.

33. Burton, G.W. Registration of Pearl Millet Inbreds Tift 23B1, Tift 23A1, Tift 23DB1, and Tift 23DA11 (Reg. Nos. PL 1, PL 2, PL 3, and PL 4). *Crop Sci.* **1969**, *9*, 397. [CrossRef]

34. Allouis, S.; Qi, X.; Lindup, S.; Gale, M.; Devos, K. Construction of a BAC library of pearl millet, *Pennisetum glaucum*. *Theor. Appl. Genet.* **2001**, *102*, 1200–1205. [CrossRef]

35. Budak, H.; Pedraza, F.; Cregan, P.; Baenziger, P.; Dweikat, I. Development and utilization of SSRs to estimate the degree of genetic relationships in a collection of pearl millet germplasm. *Crop Sci.* **2003**, *43*, 2284–2290. [CrossRef]

36. Qi, X.; Pittaway, T.; Lindup, S.; Liu, H.; Waterman, E.; Padi, F.; Hash, C.T.; Zhu, J.; Gale, M.D.; Devos, K.M. An integrated genetic map and a new set of simple sequence repeat markers for pearl millet, *Pennisetum glaucum*. *Theor. Appl. Genet.* **2004**, *109*, 1485–1493. [CrossRef]

37. Rajaram, V.; Nepolean, T.; Senthilvel, S.; Varshney, R.K.; Vadez, V.; Srivastava, R.K.; Shah, T.M.; Supriya, A.; Kumar, S.; Kumari, B.R.; et al. Pearl millet [*Pennisetum glaucum* (L.) R. Br.] consensus linkage map constructed using four RIL mapping populations and newly developed EST-SSRs. *BMC Genom.* **2013**, *14*, 159. [CrossRef]

38. Senthilvel, S.; Jayashree, B.; Mahalakshmi, V.; Kumar, P.S.; Nakka, S.; Nepolean, T.; Hash, C.T. Development and mapping of simple sequence repeat markers for pearl millet from data mining of expressed sequence tags. *BMC Plant Biol.* **2008**, *8*, 119. [CrossRef]

39. Prasanth, V.; Chandra, S.; Hoisington, D.; Jayashree, B. *AlleloBin: A Program for Allele Binning in Microsatellite Markers based on the Algorithm of Idury and Cardon (1997)*; International Crops Research Institute for the Semi-Arid Tropics (ICRISAT): New Delhi, India, 2006.

40. Elshire, R.J.; Glaubitz, J.C.; Sun, Q.; Poland, J.A.; Kawamoto, K.; Buckler, E.S.; Mitchell, S.E. A robust, simple genotyping-by-sequencing (GBS) approach for high diversity species. *PLoS ONE* **2011**, *6*, e19379. [CrossRef] [PubMed]

41. Lu, F.; Lipka, A.E.; Glaubitz, J.; Elshire, R.; Cherney, J.H.; Casler, M.D.; Buckler, E.S.; Costich, D.E. Switchgrass genomic diversity, ploidy, and evolution: Novel insights from a network-based SNP discovery protocol. *PLoS Genet.* **2013**, *9*, e1003215. [CrossRef] [PubMed]

42. Bradbury, P.J.; Zhang, Z.; Kroon, D.E.; Casstevens, T.M.; Ramdoss, Y.; Buckler, E.S. TASSEL: Software for association mapping of complex traits in diverse samples. *Bioinformatics* **2007**, *23*, 2633–2635. [CrossRef] [PubMed]

43. Bidinger, F.; Blümmel, M. Determinants of ruminant nutritional quality of pearl millet [*Pennisetum glaucum* (L.) R. Br.] stover: I. Effects of management alternatives on stover quality and productivity. *Field Crops Res.* **2007**, *103*, 119–128. [CrossRef]

44. Blümmel, M.; Bidinger, F.; Hash, C. Management and cultivar effects on ruminant nutritional quality of pearl millet (*Pennisetum glaucum* (L.) R. Br.) stover: II. Effects of cultivar choice on stover quality and productivity. *Field Crops Res.* **2007**, *103*, 129–138. [CrossRef]

45. Liu, K.; Muse, S.V. PowerMarker: An integrated analysis environment for genetic marker analysis. *Bioinformatics* **2005**, *21*, 2128–2129. [CrossRef] [PubMed]

46. Li, Y.; Guan, R.; Liu, Z.; Ma, Y.; Wang, L.; Li, L.; Lin, F.; Luan, W.; Chen, P.; Yan, Z.; et al. Genetic structure and diversity of cultivated soybean (*Glycine max* (L.) Merr.) landraces in China. *Theor. Appl. Genet.* **2008**, *117*, 857–871. [CrossRef]

47. Upadhyaya, H.D.; Dwivedi, S.L.; Baum, M.; Varshney, R.K.; Udupa, S.M.; Gowda, C.L.; Hoisington, D.; Singh, S. Genetic structure, diversity, and allelic richness in composite collection and reference set in chickpea (*Cicer arietinum* L.). *BMC Plant Biol.* **2008**, *8*, 106. [CrossRef]

48. Excoffier, L.; Smouse, P.E.; Quattro, J.M. Analysis of molecular variance inferred from metric distances among DNA haplotypes: Application to human mitochondrial DNA restriction data. *Genetics* **1992**, *131*, 479–491. [PubMed]

49. Perrier, X.; Jacquemoud-Collet, J.P. DARwin Software. Available online: https://darwin.cirad.fr/ (accessed on 11 March 2017).

50. Mantel, N. The detection of disease clustering and a generalized regression approach. *Cancer Res.* **1967**, *27*, 209–220. [PubMed]

51. R Core Team. *R: A Language and Environment for Statistical Computing*; R Foundation for Statistical Computing: Vienna, Austria, 2016. Available online: https://www.R-project.org/ (accessed on 21 July 2017).

52. SAS Institute Inc. *SAS/STAT® 14.1 User's Guide*; SAS Institute Inc.: Cary, NC, USA, 2017.

53. Kapila, R.; Yadav, R.; Plaha, P.; Rai, K.; Yadav, O.; Hash, C.; Howarth, C.J. Genetic diversity among pearl millet maintainers using microsatellite markers. *Plant Breed.* **2008**, *127*, 33–37. [CrossRef]

54. Sumanth, M.; Sumathi, P.; Vinodhana, N.; Sathya, M. Assessment of Genetic Distance Among the Inbred Lines of Pearl Millet (*Pennisetum glaucum* (L.) R. Br) Using SSR Markers. *Int. J. Biotechnol. Allied Fields* **2013**, *1*, 153–162.

55. He, Z.; Xie, F.; Chen, L.; Paz, M.A.D. Genetic diversity of tropical hybrid rice germplasm measured by molecular markers. *Rice Sci.* **2012**, *19*, 193–201. [CrossRef]

56. Filippi, C.V.; Aguirre, N.; Rivas, J.G.; Zubrzycki, J.; Puebla, A.; Cordes, D.; Moreno, M.V.; Fusari, C.M.; Alvarez, D.; Heinz, R.A.; et al. Population structure and genetic diversity characterization of a sunflower association mapping population using SSR and SNP markers. *BMC Plant Biol.* **2015**, *15*, 52. [CrossRef]

57. Hamblin, M.T.; Warburton, M.L.; Buckler, E.S. Empirical comparison of simple sequence repeats and single nucleotide polymorphisms in assessment of maize diversity and relatedness. *PLoS ONE* **2007**, *2*, e1367. [CrossRef]

58. Jones, E.; Sullivan, H.; Bhattramakki, D.; Smith, J. A comparison of simple sequence repeat and single nucleotide polymorphism marker technologies for the genotypic analysis of maize (*Zea mays* L.). *Theor. Appl. Genet.* **2007**, *115*, 361–371. [CrossRef]

59. Akmal, M.; Naeem, M.; Nasim, S.; Shakoor, A. Performance of different pearl millet genotypes under rainfed conditions. *J. Agric. Res.* **1992**, *30*, 53–58.

60. Byregowda, M. Performance of fodder bajra genotypes under rainfed conditions. *Curr. Res.-Univ. Agric. Sci.* **1990**, *19*, 128–129.

61. Mohammad, D.; Hussain, A.; Khan, S.; Bhatti, M.B. Performance of new pearl millet cultivars. *Pakistan J. Sci. Ind. R.* **1993**, *36*, 261–263.

62. Naeem, M.; Nasim, S.; Shakoor, A. Performance of new pearl millet varieties under rainfed conditions. *J. Agric. Res.* **1993**, *31*, 295–298.

63. Rai, K.N.; Blümmel, M.; Singh, A.K.; Rao, A.S. Variability and relationships among forage yield and quality traits in pearl millet. *Eur. J. Plant Sci. Biotechnol.* **2012**, *6*, 118–124.

64. Van Soest, P. Nitrogen metabolism. In *Nutritional Ecology of the Ruminant*, 2nd ed.; Comstock Publishing Associates-Cornell University Press: Ithaca, NY, USA, 1994; pp. 290–311.

65. Blümmel, M.; Rai, K. Stover quality and grain yield relationships and heterosis effects in pearl millet. *Int. Sorghum Millets Newsl.* **2003**, *44*, 141–145.

66. Hash, C.; Blummel, M.; Bidinger, F. Genotype x environment interactions in food-feed traits in pearl millet cultivars. *Int. Sorghum Millets Newsl.* **2006**, *47*, 153–157.

67. Vinayan, M.; Babu, R.; Jyothsna, T.; Zaidi, P.; Blümmel, M. A note on potential candidate genomic regions with implications for maize stover fodder quality. *Field Crops Res.* **2013**, *153*, 102–106. [CrossRef]

68. Zaidi, P.; Vinayan, M.; Blümmel, M. Genetic variability of tropical maize stover quality and the potential for genetic improvement of food-feed value in India. *Field Crops Res.* **2013**, *153*, 94–101. [CrossRef]

69. Blümmel, M.; Deshpande, S.; Kholova, J.; Vadez, V. Introgression of staygreen QLT's for concomitant improvement of food and fodder traits in Sorghum bicolor. *Field Crops Res.* **2015**, *180*, 228–237. [CrossRef]

70. Kristjanson, P.; Zerbini, E.; Rao, K. *Genetic Enhancement of Sorghum and Millet Residues Fed to Ruminants: An Ex Ante Assessment of Returns to Research*; Impact Assessment Series No. 3; ILRI (aka ILCA and ILRAD): Nairobi, Kenya, 1999.

71. Blümmel, M.; Rao, P.P. Economic value of sorghum stover traded as fodder for urban and peri-urban dairy production in Hyderabad, India. *Int. Sorghum Millets Newsl.* **2006**, *47*, 97–100.

72. Melchinger, A.E.; Lee, M.; Lamkey, K.R.; Woodman, W.L. Genetic diversity for restriction fragment length polymorphisms: Relation to estimated genetic effects in maize inbreds. *Crop Sci.* **1990**, *30*, 1033–1040. [CrossRef]

73. Charcosset, A.; Lefort-Busen, M.M.; Gallais, A. Relationship between herosis and heterozgosity at marker loci: A theoretical computation. *Theor. Appl. Genet.* **1991**, *81*, 571–575. [CrossRef] [PubMed]

Article

Effect of Species, Fertilization and Harvest Date on Microbial Composition and Mycotoxin Content in Forage

Daria Baholet [1,*], **Ivana Kolackova** [1], **Libor Kalhotka** [2], **Jiri Skladanka** [1] **and Peter Haninec** [1]

[1]   Mendel University in Brno, Faculty of AgriSciences, Department of Animal Nutrition and Forage Production, Zemedelska 3, 613 00 Brno, Czech Republic; ivana.kolackova@mendelu.cz (I.K.); jiri.skladanka@mendelu.cz (J.S.); peter.haninec@mendelu.cz (P.H.)

[2]   Department of Agrochemistry, Soil Science, Microbiology and Plant Nutrition, Zemedelska 3, 613 00 Brno, Czech Republic; libor.kalhotka@mendelu.cz

*   Correspondence: daria.baholet@mendelu.cz; Tel.: +421-918-319-326

Received: 25 March 2019; Accepted: 2 May 2019; Published: 6 May 2019

Abstract: The aim of the project was to evaluate the potential of microbial threat to feed safety in the year 2018. Analyses of the epiphytic community of several forage species (clovers, cocksfoot, fescue, festulolium, perennial ryegrass, timothy and trefoil) in variants of fertilized and non-fertilized vegetation were performed. The hypothesis is based on the fact that microorganisms are normally present on plant material during its growth all the way from the seed to the senescence; they are influenced by a plant's fitness, and they affect its harvest and utilization. Microflora was analyzed by cultivation on specific substrates, total microbial count and five specific microbial groups were observed and quantified. Forage species did not affect plant microflora. The highest risk factor of microbial contamination of feed was proved to be harvest date. Mycotoxin contamination of fresh feed was determined (deoxynivalenol and zearalenone) using ELISA. Zearalenone (ZEA) levels were negatively correlated to fertilization intensity, although these results were not statistically significant. Deoxynivalenol (DON) levels were the lowest in a moderate fertilization regime. Significant differences in mycotoxin content were found among botanical species.

Keywords: grass; clover; epiphytic microflora; fungi; deoxynivalenol; zearalenone

---

1. Introduction

Electricity consumption is increasing rapidly in the Czech Republic and decarbonization of its production is under way. One of the more sustainable production methods includes biogas stations [1]. With more than 50% of land being agricultural, possible energy sources for biogas stations are abundant [2]. This, however, presents a challenge of utilization of the biogas secondary product referred to as digestate. Digestate has the potential of becoming a new sustainable form of semi-liquid fertilizer [1].

The surface above ground biomass, such as leaves, stems or reproduction organs, is called phyllosphere [3]. It is colonized by a wide array of microorganisms; approximately 37 bacterial and 12 fungal genera is present on the wheat's leaves [4]. Microbial composition of the surface is affected by weather conditions and geographical location [5].

Microbial contamination of animal feed is closely related to research of nutritional pathology and shows its consequent effects on animal susceptibility to disease and use of antibiotics in livestock breeding [6]. Economic loss due to fungal pathogens is an important factor in feed production, with the main genera causing decreased yield and quality are *Fusarium*, *Aspergillus* and *Penicillium* [7]. Silage is a fermented feed highly used in temperate areas since the 1960s and bacterial species play a fundamental role in its production [8]. Besides lactic acid bacteria, used as a starter additive for silaging, there are species responsible for spoilage and potential health-deteriorative properties [9].

Many species of filamentous fungi produce secondary metabolites harmful to vertebrates. Also belonging to this category, are mycotoxins, frequently occurring in cereals and other feedstuffs [10]. Mycotoxins are produced by molds under specific conditions and their production is promoted in high humidity, poor agricultural practices (e.g., inadequate fertilization, disuse of crop rotation, contaminated seeds) or damaged and contaminated crops. Although the presence of molds on grains does not necessarily mean there are mycotoxins present, the potential for mycotoxin production does exist. Furthermore, the long-term absence of molds on stored food and feed does not guarantee that the grain is free of mycotoxins [11]. The issue of mycotoxin risk is, therefore, tricky and requires the attention of both agrotechnology specifically and the scientific community generally.

For food and feed safety, the most notable mycotoxins are aflatoxins, ochratoxin A, fumonisins, deoxynivalenol, patulin, and zearalenone [12]. However, there are more than 500 mycotoxins known nowadays [13]. Frequently, more than one mycotoxin contaminates feed [14]. Production depends on various factors, including temperature, water activity and genotype of the mycotoxigenic species [15,16]. In silaging, there is clear evidence of fungal and consecutive mycotoxin production inhibition by lactic acid bacteria in the fermentation process [17,18].

There is not sufficient amount of information dedicated to above-ground biomass of grass and clover species, so it is essential to broaden the knowledge on effects of environmental conditions, microbial composition and plant cultivars of the Czech Republic.

2. Materials and Methods

2.1. Experimental Plot Maintenance

This study was conducted on seven forage species and their cultivars: ×*Festulolium* 'Felina', *Dactylis glomerata* L. 'Vega', *Festuca arundinacea* Schreb. 'Prosteva', *Lolium perenne* L. 'Promed' and 'Proly', *Phleum pratense* L. 'Sobol', *Trifolium pratense* L. 'Spurt' and 'Blizard', *Trifolium repens* L. 'Klondike'.

Plots were located in Research Station Vatín, Czech Republic (49°52′ N, 15°96′ E) situated 560 m above sea level, with annual precipitation of 617 mm and mean annual temperature of 6.9 °C. Soil type of the chosen experimental location was Cambisol as a sandy-loam soil on the diluvium of biotic orthognesis. Analyses were conducted in Brno, Czech Republic (49°21′ N, 16°61′ E). A randomized plot design was used and three repetitions of each variant was sown in small plots (1.25 × 8 m).

Spring fertilization of the experimental field was done by digestate. Input material for digestate provided by biogas station Pikarec was maize silage (13,210 tons were processed in biogas station per year 2018), cow manure (3000 t/year), cereal silage (2000 t/year), grass silage (1000 t/year) and fresh grass forage (500 t/year). Details on chemical composition of the digestate are described in Table 1; pH value was 7.77.

Table 1. Chemical composition of digestate (measured in 100 g of digestate).

Nutrient	Content (%)
Dry matter	5.70
Total nitrogen	0.49
Phosphorus	0.06
Potassium	0.42
Calcium	0.13
Magnesium	0.05

The digestate was applied in October 2017 and after each harvest. Mixed samples were created by blending of samples gathered from fertilization variants A, B and C for purposes of creating reference sample. Fertilization regime on the said three variants of plots were as described below.

- Variant A: non-fertilized control
- Variant B (150 kg N/ha/year): Total amount of nitrogen was applied in three doses, therefore each application dose contained 1/3 of annual dose = 9.43 kg digestate per 10m^2 (equal to 50 kg N/ha/application)
- Variant C (300 kg N/ha/year): Total amount of nitrogen was applied in three doses, therefore each application dose contained 1/3 of annual dose = 18.87 kg digestate per 10m^2 (equal to 100 kg N/ha/application)

Fresh biomass was collected in two harvests: 17 May 2018 and 14 July 2018, subsequently chopped into 3 cm pieces, chilled and immediately transported to specialized laboratory in Brno.

2.2. Microbial Analyses

Sample (10 g) of the original fresh biomass or silage was shaken on a PSU-10i orbital shaker (Biosan, Riga, Latvia) for 10 min with 90 mL of sterile saline. A series of ten-fold dilutions were then prepared from the solution. These groups of microorganisms were determined in the samples after cultivation as follows:

- Total microbial count (TMC) on Plate Count Agar (Biokar Diagnostics, Pantin, France) at 30 °C for 72 h.
- Lactic acid bacteria (LAB) on De Man, Rogosa and Sharpe Agar (Biokar Diagnostics, Pantin, France) at 30 °C for 72 h.
- Enterococcus sp. on COMPASS Enterococcus agar (Biokar Diagnostics, Pantin, France) at 44 °C for 24 h.
- Enterobacteriaceae on Violet Red Bile Glucose Agar (Biokar Diagnostics, Pantin, France) at 37°C for 24 h.
- Micromycetes (yeasts and molds) on Chloramphenicol Glucose Agar (Biokar Diagnostics, Pantin, France) at 25 °C for 120 h.

After the cultivation time, CFUs were counted on ColonyStar colony counter (Funke Gerber, Berlin, Germany) equipped with pressure-sensitive automatic counter and illuminated counting plate. The result was expressed as a number of colony-forming units per gram of sample.

2.3. Mycotoxin Analyses

Samples from species ×Festulolium 'Felina', *Festuca arundinacea* L. 'Prosteva' *Lolium perenne* L. 'Promed' and *Phleum pratense* L. 'Sobol' were prepared by drying fresh biomass at 60 °C. Dried samples were milled to 1 mm particles (Pulverisette laboratory cutting mill; Fritsch, Weimar, Germany) and supernatant was created for further testing by ELISA method. For DON 2 g of milled homogenate were weighted and 20 mL of distillated water was added. In ZEA analysis 2 g of milled sample homogenate were weighted and 8 mL of 90% methyl alcohol was added.

An ELISA method was applied for estimation of the mycotoxin contents. The ELISA assay test was a competitive direct enzyme-linked immunosorbent assay used for the quantitative analysis of mycotoxins. The test kits (MyBioSource, San Diego, CA, USA) were provided in a microwell format. The test was read in a microwell reader. The optical densities of the control formed the standard curve, and the sample optical densities were plotted against the curve to calculate the exact concentration of toxins [19]. Wavelenght in Synergy HTX Multi-Mode Microplate Reader (BioTek, Winooski, VT, USA) was adjusted to 450 nm and mycotoxin content was determined.

2.4. Statistical Analyses

Data were evaluated using StatSoft Statistica 12.0 (TIBCO Software Inc., Palo Alto, CA, USA). Data was tested for normality of distribution by Shapiro–Wilk test. Kruskal–Wallis ANOVA analysis was conducted with microbial data, single-factor ANOVA and Scheffé test were used in mycotoxin data. Significant differences were accepted if $p < 0.05$.

3. Results

3.1. Date of Sampling

The date of sampling does not affect the TMC, but it influences the composition of the microbiome of the plant. No statistical differences were found between the May and July collection dates. TMC samples showed a trend of decreasing microbial counts later in the season. Similarly, there was an increase in micromycetes and also yeasts in later sampling dates. Mean values and statistical differences are summarized in Table 2.

Table 2. Date of sampling effects on microbial counts in plant samples.

Date of Sampling	Microbial Group (CFU/g)	Mean Value	SE	p	H
17.5.2018	Total microbial count	4.39×10^5	1.95×10^8	0.2372 [a]	1.3974
	Lactic acid bacteria	3.83×10^5	1.50×10^5	0.7874 [a]	0.0727
	Enterococcus sp.	5.96×10^5	2.22×10^5	0.5331 [a]	0.3884
	Enterobacteriaceae	4.44×10^5	1.65×10^5	0.0727 [a]	0.7874
	Total micromycetes	6.14×10^5	3.01×10^5	0.0712 [a]	3.2562
	Yeasts	6.00×10^5	3.00×10^5	0.7557 [a]	0.0967
	Filamentous fungi	1.27×10^4	3.29×10^3	0.0001 [b]	18.7958
14.7.2018	Total microbial count	2.27×10^8	1.26×10^8	0.2372 [a]	1.3974
	Lactic acid bacteria	3.20×10^3	8.81×10^2	0.7874 [a]	0.0727
	Enterococcus sp.	8.86×10^2	5.28×10^2	0.5331 [a]	0.3884
	Enterobacteriaceae	1.53×10^4	5.66×10^3	0.0727 [a]	0.7874
	Total micromycetes	1.60×10^6	9.00×10^5	0.0712 [a]	3.2562
	Yeasts	1.33×10^6	9.14×10^5	0.7557 [a]	0.0967
	Filamentous fungi	2.61×10^5	7.79×10^4	0.0001 [b]	18.7958

SE shows values of standard error of the mean. H shows results of Kruskal–Wallis test. Mean values are statistically significant in $p < 0.05$. These values are marked with a different letter in the upper index.

Statistically significant differences between May and July sample collection dates were found only in filamentous fungi. In May, we measured an amount that was significantly lower (1.27×10^4 CFU/g) compared to July. Moreover, in the second cut grassland, fungal counts increased by more than 100% (2.61×10^5 CFU/g).

3.2. Botanical Species

Statistically significant effects of botanical species on microbial communities of epiphytic plant sections were not observed in our experiment. Highly variable data were gathered in the context of plant species. When assessed from a practical standpoint, it is important to note that higher filamentous fungi count and a, thereby, likely higher in mycotoxin contamination risk was observed in *Lolium perenne* L. 'Proly'.

3.3. Fertilization

The trend of highest microbial counts in fertilization extremes (either unfertilized control or highly fertilized variant C) was observed also in other microbial groups (Table 3). There were high deviations from mean values measured, however, data indicate the optimal fertilization regime is variant B despite the influence of other factors. Especially, amounts of enterococci were minimal in moderate fertilization regime, mean *Enterococcus* sp. was 81 CFU/g. The contrary was visible in case of filamentous fungi, highest values were obtained from a moderately fertilized plot (2.88×10^5 CFU/g). Significant differences were only found among fertilization variants and mixed sample, however mixed sample results were only included as a benchmark for values.

Table 3. Fertilization effects on microbial counts in plant samples.

Fertilization Regime	Microbial Group (CFU/g)	Mean Value	SE	*p*	H
Variant A (0 kg N/ha)	Total microbial count	3.50×10^8	2.41×10^8	0.0001 [a]	20.4679
	Lactic acid bacteria	5.20×10^5	2.65×10^5	0.0040 [a]	13.3123
	Enterococcus sp.	6.86×10^5	3.60×10^5	0.0413 [a]	8.2419
	Enterobacteriaceae	5.05×10^5	2.71×10^5	0.0116 [a]	11.0197
	Total micromycetes	2.24×10^6	1.60×10^6	0.0005 [a]	17.5514
	Yeasts	2.18×10^6	1.59×10^6	0.0035 [a]	13.5785
	Filamentous fungi	6.27×10^4	1.62×10^4	0.0025 [a]	14.3612
Variant B (150 kg N/ha)	Total microbial count	9.53×10^7	3.60×10^7	0.0001 [a]	20.4679
	Lactic acid bacteria	2.62×10^3	9.51×10^2	0.0040 [a]	13.3123
	Enterococcus sp.	81.4	30.5	0.0413 [a]	8.2419
	Enterobacteriaceae	9.72×10^3	3.62×10^3	0.0116 [a]	11.0197
	Total micromycetes	3.54×10^5	1.62×10^5	0.0005 [a]	17.5514
	Yeasts	3.64×10^4	2.62×10^4	0.0035 [a]	13.5785
	Filamentous fungi	2.88×10^5	1.51×10^5	0.0025 [a]	14.3612
Variant C (300 kg N/ha)	Total microbial count	8.34×10^8	3.04×10^8	0.0001 [a]	20.4679
	Lactic acid bacteria	1.99×10^5	9.66×10^4	0.0040 [a]	13.3123
	Enterococcus sp.	4.32×10^5	2.37×10^5	0.0413 [a]	8.2419
	Enterobacteriaceae	3.47×10^5	1.69×10^5	0.0116 [a]	11.0197
	Total micromycetes	1.65×10^6	6.82×10^5	0.0005 [a]	17.5514
	Yeasts	1.39×10^6	7.33×10^5	0.0035 [a]	13.5785
	Filamentous fungi	2.60×10^5	1.22×10^5	0.0025 [a]	14.3612

SE shows values of standard error of the mean. H shows results of Kruskal–Wallis test. Mean values are statistically significant in $p < 0.05$. These values are marked with a different letter in the upper index.

3.4. Mycotoxin Contamination of Fresh Feed

Species and fertilization were tested for affecting the mycotoxin occurrence in plant samples. Zearalenone levels in fresh biomass were generally lower than DON. Practical significance of fertilization related to mycotoxin production was found, although statistically there were no differences proven (Table 4). Both mycotoxins occurred in all fertilization regimes. DON concentrations were lowest in moderately fertilized variant B (5.03 ng/mL). Highest DON contamination was observed the highly fertilized variant C (5.32 ng/mL). ZEA concentration slightly increased with decreased use of digestate, with lowest value of 1.18 ng/mL and highest of 1.39 ng/mL (Table 4).

Table 4. Fertilization effects on mycotoxin concentration in plant samples.

Fertilization Regime	Mycotoxin Concentration (ng/mL)	Mean Value	SE	*p*
Variant A (0 kg N/ha)	Deoxynivalenol	5.0979	1.29	0.4437 [a]
	Zearalenon	1.3984	0.23	0.5574 [a]
Variant B (150 kg N/ha)	Deoxynivalenol	5.0317	1.86	0.4437 [a]
	Zearalenon	1.1825	0.11	0.5574 [a]
Variant C (300 kg N/ha)	Deoxynivalenol	5.3212	1.79	0.4437 [a]
	Zearalenon	1.1820	0.10	0.5574 [a]

SE shows values of standard error of the mean. Mean values are statistically significant in $p < 0.05$. These values are marked with a different letter in the upper index.

There were notable differences in mycotoxin contamination among botanical species (Table 5). Highest levels of DON were found in *Festuca arundinacea* (7.93 ng/mL) and *Phleum pratense* (7.63 ng/mL). Less susceptible to mycotoxin contamination were proved to be *Festulolium* (1.05 ng/mL) and *Lolium perenne* (2.43 ng/mL). Significant differences were found between these two groups, however, there were no statistical differences between ×*Festulolium* and *Lolium* or *Phleum* and *Festuca*.

Table 5. Botanical species effects on mycotoxin concentration in plant samples.

Species	Mycotoxin Concentration (ng/mL)	Mean Value	SE	p
×*Festulolium* 'Felina'	Deoxynivalenol	1.0520	0.52	0.0008 [a]
	Zearalenon	0.9665	0.03	0.4437 [b]
Festuca arundinacea L. 'Prosteva'	Deoxynivalenol	7.9331	0.91	0.0008 [a]
	Zearalenon	1.2893	0.11	0.4437 [b]
Lolium perenne L. 'Promed'	Deoxynivalenol	2.4330	1.43	0.0008 [a]
	Zearalenon	1.3480	0.37	0.4437 [b]
Phleum pratense L. 'Sobol'	Deoxynivalenol	7.6381	0.81	0.0008 [a]
	Zearalenon	1.3209	0.06	0.4437 [b]

SE shows values of standard error of the mean. Mean values are statistically significant in $p < 0.05$. These values are marked with a different letter in the upper index.

4. Discussion

Presence of microbial families and species is geographically specific, which leads us to believe that the same is true for mycotoxins [20]. Study of epiphytic microbial colonies and elements affecting their occurrence is multi-faceted, with high variability due to many apparent factors, such as climatic conditions (e.g., temperature, water activity), microbial tolerance to pH and phytochemicals or mycotoxigenic species genotype [15,16,21,22].

Not many authors have studied epiphytic plant microbiomes, and focus of microbial observations in relation to grassland management techniques is mainly on soil microbiota [23]. Moreover, there are not enough complex studies (including both bacterial and fungal species) done in grassland ecosystems and forage species. Microbial observations are mainly focused on pathogenic *Fusarium* sp. [24]. However, some authors have done experiments on bacterial communities on maize or rice [25,26]. From results gathered in this experimental study, it is possible to conclude that botanical species do not affect microbial composition of a plant's phyllosphere. Species variants with the same climatic and soil factors were observed to have similar CFU counts, which is probably true for the unmonitored weed species as well. However, this conclusion is only applicable in terms of the experimental location of this study and similar weather conditions.

In case of differences between harvest dates, statistical significance between May and July harvests was observed in fungal counts. Lower amounts were found in May, although they were abundant in July samples. It may have been caused by the average monthly air temperature of 18.3 °C in the research location, which has a positive effect on fungal growth and reproduction [27]. This may have also been caused by high pathogenic pressure of the year 2018.

Karlsson et al. [28] had previously recorded a positive correlation between production intensification (increased nitrogen concentrations by fertilization) and higher microbial counts. We found no significant difference between microbial counts in any of the fertilization regimes.

In late autumn, usually, the vegetation of pasture plants gradually decreases and weather conditions stimulate the development of microscopic fungi which, in consequence, may lead to the formation of mycotoxins [29–31]. Besides population density, the formation of mycotoxins additionally depends on several biotic and abiotic factors [32,33]. These metabolites can cause economic losses in animal production and decrease meat quality [34].

DON has an important physicochemical ability of withstanding high temperatures, which increases the risk of its occurrence in food [35]. In another research, it was analyzed that in animals that were exposed to the (DON) a subsequent transfer of this toxin to animal products was found. However, the rate of transmission was low. Overall, the study showed that short-term and sub-chronic exposure to DON decreased body weight, weight gain, and feed consumption in rats and mice. Haematological effects were also observed [36].

Zearalenone is one of mycotoxins produced by the *Fusarium* genus. It can be detected in the forage of grass stands. It exhibits high oestrogenic activity. Apart from the direct impact on ruminants, the contaminated forage affects rumen microorganisms, too [37]. Forage with a zearalenone content higher than 0.5 mg/kg is not advised for feeding [38].

Many strategies can be employed in the process of decreasing the mycotoxin contamination of animal feed, the most effective is prevention of contamination on the field. One of these strategies may be ensuring optimal nutrient content in soil. There was a visible decreasing trend of ZEA contamination with an increase of digestate amount. From this negative correlation, we can conclude that higher levels of fertilization can help in prevention from fungal degradation of feed in case of ZEA. This, however is not the most efficient way to combat DON accrual in feed. Therefore, the moderate intensity of fertilization is an optimal solution.

The contents of DON and ZEA depended on the course of weather, too. Sutton et al. [39] stated, that rainfall increases the occurrence of zearalenone in corn during summer, although different temperatures did not have an effect on its production. During the growing season, forage grasses may become contaminated with mycotoxins. This phenomenon mainly occurred in May and in July, which means a high risk of mycotoxin input to the food chain.

When cattle are grazing in winter, a higher occurrence of mycotoxins in the feed may be expected. Related damage to animal metabolism may affect the number of diseased animals and/or diagnostics of animal diseases. Consequently, mycotoxins impact not only performance and health of animals, but also overall economy of production [40]. However, some authors estimate that breeding of new species and improvement of currently used forage lines appears to be the most perspective approach [41,42]. From results of our study, we can conclude that in climatic and soil conditions of Vatin (Czech Republic) ×*Festulolium* hybrids appear to be promising due to their lower mycotoxin contamination. Post-harvest technologies have been studied heavily in recent years, including use of essential oils, LAB additives or acidic electrolyzed water [43–45].

Author Contributions: Responsible for conceptualization, methodology, writing—review and editing are J.S. and I.K.; Software: P.H.; Validation and supervision: J.S.; Formal analysis: P.H.; Investigation, funding acquisition, project administration: D.B. Data curation: D.B. and L.K.; Writing—original draft preparation: I.K. and D.B.; Visualization: D.B. and I.K.; Resources: Internal Grant Agency.

Funding: This research was funded by Internal Grant Agency of Mendel University in Brno, grant number AF-IGA-IP-2018/028. The APC was funded by AF-IGA-IP-2018/028 as well.

Conflicts of Interest: The authors declare no conflict of interest. The funders had no role in the design of the study; in the collection, analyses, or interpretation of data; in the writing of the manuscript, or in the decision to publish the results.

References

1. Simeckova, J.; Jandak, J.; Masicek, T. *Vliv Aplikace Digestátu na Vlastnosti Kambizemě*; Mendelova Univerzita v Brně: Brno, Czech Republic, 2018; ISBN 978-80-7509-613-5.
2. Agriculture in the Czech Republic. Embassy of the Czech Republic in Tel Aviv. Available online: https://www.mzv.cz/telaviv/en/economy_and_trade/agriculture_in_the_czech_republic/ (accessed on 17 April 2019).
3. Thapa, S.; Prasanna, R.; Ranjan, K.; Velmourougane, K.; Ramakrishnan, B. Nutrients and host attributes modulate the abundance and functional traits of phyllosphere microbiome in rice. *Microbiol. Res.* **2017**, *204*, 55–64. [CrossRef]
4. Legard, D.E.; McQuilken, M.P.; Whipps, J.M.; Fenlon, J.S.; Fermor, T.R.; Thompson, I.P.; Bailey, M.J.; Lynch, J.M. Studies of seasonal changes in the microbial populations on the phyllosphere of spring wheat as a prelude to the release of a genetically modified microorganism. *Agric. Ecosyst. Environ.* **1994**, *50*, 87–101. [CrossRef]
5. Nugmanov, A.; Beishova, I.; Kokanov, S.; Lozowicka, B.; Kaczynski, P.; Konecki, R.; Snarska, K.; Wołejko, E.; Sarsembayeva, N.; Abdigaliyeva, T. Systems to reduce mycotoxin contamination of cereals in the agricultural region of Poland and Kazakhstan. *Crop Prot.* **2018**, *106*, 64–71. [CrossRef]
6. Ravindran, V. Nutrition and pathology of non-ruminants. *Anim. Feed Sci. Technol.* **2012**, *173*, 1–2. [CrossRef]
7. Nguyen, P.-A.; Strub, C.; Fontana, A.; Schorr-Galindo, S. Crop molds and mycotoxins: Alternative management using biocontrol. *Biol. Control* **2017**, *104*, 10–27. [CrossRef]

8. Cheli, F.; Campagnoli, A.; Dell'Orto, V. Fungal populations and mycotoxins in silages: From occurrence to analysis. *Anim. Feed Sci. Technol.* **2013**, *183*, 1–16. [CrossRef]

9. Ogunade, I.M.; Jiang, Y.; Pech Cervantes, A.A.; Kim, D.H.; Oliveira, A.S.; Vyas, D.; Weinberg, Z.G.; Jeong, K.C.; Adesogan, A.T. Bacterial diversity and composition of alfalfa silage as analyzed by Illumina MiSeq sequencing: Effects of *Escherichia coli* O157:H7 and silage additives. *J. Dairy Sci.* **2018**, *101*, 2048–2059. [CrossRef] [PubMed]

10. Bezerra da Rocha, M.E.; da Chagas Oliveira FreireFreire, F.; Maia, F.E.F.; Guedes, M.I.F.; Rondina, D. Mycotoxins and their effects on human and animal health. *Food Control* **2014**, *36*, 159–165. [CrossRef]

11. Edwards, S.G. Influence of agricultural practices on fusarium infection of cereals and subsequent contamination of grain by trichothecene mycotoxins. *Toxicol. Lett.* **2004**, *153*, 29–35. [CrossRef]

12. Pitt, J.I. Mycotoxins: Mycotoxins—General. In *Encyclopedia of Food Safety*; Motarjemi, Y., Ed.; Academic Press: Waltham, MA, USA, 2014; pp. 283–288. ISBN 978-0-12-378613-5.

13. Stein, R.A.; Bulboaca, A.E. Chapter 21—Mycotoxins. In *Foodborne Diseases*, 3rd ed.; Dodd, C.E.R., Aldsworth, T., Stein, R.A., Cliver, D.O., Riemann, H.P., Eds.; Academic Press: Cambridge, MA, USA, 2017; pp. 407–446. ISBN 978-0-12-385007-2.

14. Manizan, A.L.; Oplatowska-Stachowiak, M.; Piro-Metayer, I.; Campbell, K.; Koffi-Nevry, R.; Elliott, C.; Akaki, D.; Montet, D.; Brabet, C. Multi-mycotoxin determination in rice, maize and peanut products most consumed in Côte d'Ivoire by UHPLC-MS/MS. *Food Control* **2018**, *87*, 22–30. [CrossRef]

15. Mateo, E.M.; Valle-Algarra, F.M.; Jimenez, M.; Magan, N. Impact of three sterol-biosynthesis inhibitors on growth of *Fusarium langsethiae* and on T-2 and HT-2 toxin production in oat grain under different ecological conditions. *Food Control* **2013**, *34*, 521–529. [CrossRef]

16. Aldars-Garcia, L.; Berman, M.; Ortiz, J.; Ramos, A.J.; Marin, S. Probability models for growth and aflatoxin B1 production as affected by intraspecies variability in *Aspergillus flavus*. *Food Microbiol.* **2018**, *72*, 166–175. [CrossRef]

17. Quattrini, M.; Bernardi, C.; Stuknytė, M.; Masotti, F.; Passera, A.; Ricci, G.; Vallone, L.; De Noni, I.; Brasca, M.; Fortina, M.G. Functional characterization of *Lactobacillus plantarum* ITEM 17215: A potential biocontrol agent of fungi with plant growth promoting traits, able to enhance the nutritional value of cereal products. *Food Res. Int.* **2018**, *106*, 936–944. [CrossRef]

18. Luz, C.; Ferrer, J.; Mañes, J.; Meca, G. Toxicity reduction of ochratoxin A by lactic acid bacteria. *Food Chem. Toxicol.* **2018**, *112*, 60–66. [CrossRef]

19. Skladanka, J.; Nedenik, J.; Adam, V.; Dolezal, P.; Moravcova, H.; Dohnal, V. Forage as a Primary Source of Mycotoxins in Animal Diets. *Int. J. Environ. Res. Public Health* **2010**, *8*, 37–50. [CrossRef] [PubMed]

20. Chen, W.; Turkington, T.K.; Lévesque, C.A.; Bamforth, J.M.; Patrick, S.K.; Lewis, C.T.; Chapados, J.T.; Gaba, D.; Tittlemier, S.A.; MacLeod, A.; et al. Geography and agronomical practices drive diversification of the epiphytic mycoflora associated with barley and its malt end product in western Canada. *Agric. Ecosyst. Environ.* **2016**, *226*, 43–55. [CrossRef]

21. Maciorowski, K.G.; Herrera, P.; Jones, F.T.; Pillai, S.D.; Ricke, S.C. Effects on poultry and livestock of feed contamination with bacteria and fungi. *Anim. Feed Sci. Technol.* **2007**, *133*, 109–136. [CrossRef]

22. Yadav, R.K.P.; Papatheodorou, E.M.; Karamanoli, K.; Constantinidou, H.-I.A.; Vokou, D. Abundance and diversity of the phyllosphere bacterial communities of Mediterranean perennial plants that differ in leaf chemistry. *Chemoecology* **2008**, *18*, 217–226. [CrossRef]

23. Ma, B.; Cai, Y.; Bork, E.W.; Chang, S.X. Defoliation intensity and elevated precipitation effects on microbiome and interactome depend on site type in northern mixed-grass prairie. *Soil Biol. Biochem.* **2018**, *122*, 163–172. [CrossRef]

24. Morcia, C.; Rattotti, E.; Stanca, A.M.; Tumino, G.; Rossi, V.; Ravaglia, S.; Germeier, C.U.; Herrmann, M.; Polisenska, I.; Terzi, V. Fusarium genetic traceability: Role for mycotoxin control in small grain cereals agro-food chains. *J. Cereal Sci.* **2013**, *57*, 175–182. [CrossRef]

25. Kharazian, Z.A.; Salehi Jouzani, G.; Aghdasi, M.; Khorvash, M.; Zamani, M.; Mohammadzadeh, H. Biocontrol potential of *Lactobacillus* strains isolated from corn silages against some plant pathogenic fungi. *Biol. Control* **2017**, *110*, 33–43. [CrossRef]

26. Heron, S.J.E.; Wilkinson, J.F.; Duffus, C.M. Enterobacteria associated with grass and silages. *J. Appl. Bacteriol.* **1993**, *75*, 13–17. [CrossRef]

27. Kozlovska, S.; Zahradnicek, P. INFOMET | Informační web ČHMÚ | Český Hydrometeorologický Ústav | Meteorologie, Klimatologie, Hydrologie, Čistota Ovzduší, Předpověď Počasí. Available online: http://www.infomet.cz/index.php?id=read&idd=1503567031 (accessed on 6 March 2019).

28. Karlsson, I.; Friberg, H.; Kolseth, A.; Steinberg, C.; Persson, P. Agricultural factors affecting *Fusarium* communities in wheat kernels. *Int. J. Food Microbiol.* **2017**, *252*, 53–60. [CrossRef]

29. Giesler, L.J.; Yuen, G.Y.; Horst, G.L. The microclimate in tall fescue turf as affected by canopy density and its influence on brown patch disease. *Plant Dis.* **1996**, *80*, 389–394. [CrossRef]

30. Opitz von Boberfeld, W.; Banzhaf, K. Yield and forage quality of different x *Festulolium* cultivars in winter. *J. Agron. Crop Sci.* **2006**, *192*, 239–247. [CrossRef]

31. Behrendt, U.; Stauber, T.; Müller, T. Microbial communities in the phyllosphere of grasses on fenland at different intensities of management. *Grass Forage Sci.* **2004**, *59*, 169–179. [CrossRef]

32. De Nijs, M.; Soentoro, P.; Delfgou-Van Asch, E.; Kamphuis, H.; Rombouts, F.M. Fungal infection and presence of deoxynivalenol and zearalenone in The Netherlands. *J. Food Prot.* **1996**, *59*, 772–777. [CrossRef]

33. Engels, R.; Krämer, J. Incidence of fusaria and occurrence of selected *Fusarium* mycotoxins on *Lolium* ssp. in Germany. *Mycotoxin Res.* **1996**, *12*, 31–40. [CrossRef]

34. Opitz von Boberfeld, W. Changes of the quality including mycotoxin problems of the primary growth of a hay meadow—*Arrhenatherion elatioris*. *Agribiol. Res.* **1996**, *49*, 52–62.

35. Sobrova, P.; Adam, V.; Vasatkova, A.; Beklova, M.; Zeman, L.; Kizek, R. Deoxynivalenol and its toxicity. *Interdiscip. Toxicol.* **2010**, *3*, 94–99. [CrossRef]

36. Hughes, D.M.; Gahl, M.J.; Graham, C.H.; Grieb, S.L. Overt signs of toxicity to dogs and cats of dietary deoxynivalenol. *J. Anim. Sci.* **1999**, *77*, 693–700. [CrossRef]

37. Wolf, D. On the Effect of Stand, Pre-Utilization and Date of Winter Harvest on Quality and Yield of Winter Pasture. Ph.D. Thesis, Justus Liebig University, Giessen, Germany, 2002.

38. Marasas, W.F.O.; Van Rensburg, S.J.; Mirocha, C.J. Incidence of fusarium species and the mycotoxins, deoxynivalenol and zearalenone, in corn produced in esophageal cancer areas in Transkei. *J. Agric. Food Chem.* **1979**, *27*, 1108–1112. [CrossRef]

39. Sutton, J.C.; Baliko, W.; Funnell, H.S. Relation of Weather Variables to Incidence of Zearalenone in Corn in Southern Ontario. *Can. J. Plant Sci.* **1980**, *60*, 149–155. [CrossRef]

40. Skladanka, J.; Dohnal, V.; Dolezal, P.; Jezkova, A.; Zeman, L. Factors Affecting the content of Ergosterol and Zearalenone in Selected Grass Species at the End of the Growing season. *Acta Vet. Brno* **2009**, *78*, 353–360. [CrossRef]

41. Peng, W.-X.; Marchal, J.L.M.; van der Poel, A.F.B. Strategies to prevent and reduce mycotoxins for compound feed manufacturing. *Anim. Feed Sci. Technol.* **2018**, *237*, 129–153. [CrossRef]

42. Wiwart, M.; Suchowilska, E.; Kandler, W.; Sulyok, M.; Groenwald, P.; Krska, R. Can Polish wheat (*Triticum polonicum* L.) be an interesting gene source for breeding wheat cultivars with increased resistance to *Fusarium* head blight? *Genet. Resour. Crop Evol.* **2013**, *60*, 2359–2373. [CrossRef]

43. Xu, L.; Tao, N.; Yang, W.; Jing, G. Cinnamaldehyde damaged the cell membrane of *Alternaria alternata* and induced the degradation of mycotoxins in vivo. *Ind. Crops Prod.* **2018**, *112*, 427–433. [CrossRef]

44. Ma, Z.X.; Amaro, F.X.; Romero, J.J.; Pereira, O.G.; Jeong, K.C.; Adesogan, A.T. The capacity of silage inoculant bacteria to bind aflatoxin B1 in vitro and in artificially contaminated corn silage. *J. Dairy Sci.* **2017**, *100*, 7198–7210. [CrossRef] [PubMed]

45. Lyu, F.; Gao, F.; Zhou, X.; Zhang, J.; Ding, Y. Using acid and alkaline electrolyzed water to reduce deoxynivalenol and mycological contaminations in wheat grains. *Food Control* **2018**, *88*, 98–104. [CrossRef]

![agriculture logo] *agriculture*

[MDPI]

Article

The Effect of Herbage Conservation Method on Protein Value and Nitrogen Utilization in Dairy Cows

Christian Böttger [1], Paolo Silacci [2], Frigga Dohme-Meier [3], Karl-Heinz Südekum [1] and Ueli Wyss [3,*]

[1] Institute of Animal Science, University of Bonn, Endenicher Allee 15, 53115 Bonn, Germany; cboe@itw.uni-bonn.de (C.B.); ksue@itw.uni-bonn.de (K.-H.S.)
[2] Agroscope, Animal Biology Research Group, Tioleyre 4, 1725 Posieux, Switzerland; paolo.silacci@agroscope.admin.ch
[3] Agroscope, Ruminant Research Unit, Tioleyre 4, 1725 Posieux, Switzerland; frigga.dohme-meier@agroscope.admin.ch
* Correspondence: ueli.wyss@agroscope.admin.ch; Tel.: +41-58-466-7214

Received: 3 May 2019; Accepted: 2 June 2019; Published: 6 June 2019

Abstract: Ruminant production systems frequently rely on grassland utilization and conservation of herbage as hay or silage. Conservation affects the crude protein (CP) composition and protein value, which is particularly recognized during ensiling. The aim of the current study was to describe the effect of the conservation method on forage protein value and N utilization in dairy cows. Herbage from the same sward was cut and conserved as silage (SI), barn-dried hay (BH), or field-dried hay (FH). Laboratory evaluation indicated differences in CP fractions and ruminal degradability of CP. Conserved forages were fed to six lactating Holstein cows in a replicated 3 × 3 Latin square design, and N balance was assessed. Partitioning of N into milk, feces, and urine was affected only moderately. Lower concentrations of serum, milk, and also urinary urea indicated lower N turnover for FH compared to SI and BH, likely due to lower N intake for FH. However, the use efficiency of feed N for milk N did not differ between the types of forage. Further, high CP concentrations and the unbalanced concentrations of CP and energy in the forages led to excess excretion of N in all treatments and presumably superimposed effects of the conservation method on N utilization.

Keywords: digestibility; energy balance; forage; hay; nitrogen balance; silage

1. Introduction

Worldwide, agriculture substantially relies on grassland utilization. Feeding high amounts of forage to ruminants is beneficial with regard to maintaining rumen function and reduced competition with resources for human nutrition [1]. Moreover, utilization of forage produced on farm can be advantageous over imported concentrate in terms of both cost and nutrient cycles.

In many countries, conservation of herbage plays a key role, either to supply forage for winter feeding or as year-round feed in stall-feeding systems. Ensiling is often favored to conserve herbage in humid and temperate regions due to a reduced period between cutting and harvesting [2]. However, traditional conservation as hay has gained renewed interest in grassland-dominated regions specialized in the production of dairy products with different quality labels, such as protected-designation-of-origin (PDO) hard cheese types (e.g., Gruyère cheese [3]). These labels often offer higher milk payment but prohibit the feeding of silage, e.g., because of concerns regarding lowered cheese processing quality caused by clostridia contamination [4]. Haymaking in the field requires constant weather conditions for several days, which causes some uncertainty for the production of high-quality forage. A way to reduce the time in the field is conservation as barn-dried hay where the fresh herbage is put on a ventilation just after wilting in the field for some days [5].

However, conservation of forages—and particularly, ensiling—can have significant effects on crude protein (CP) composition of the forage and N utilization by the animal [6,7]. This is mainly due to the fact that much of the original true protein (TP) is degraded to non-protein N (NPN) during ensiling. Crude protein degradation in dried forages is generally less pronounced than in silages [8]. True protein concentration as an indicator of protein degradation during conservation is routinely included in silage quality evaluation by many laboratories. However, a more detailed fractionation of feed CP according to the Cornell Net Carbohydrate and Protein System (CNCPS [9,10]) could provide a better understanding of the effect of the conservation method on herbage quality. The distribution of CP fractions per se can reveal potential conservation-induced changes in herbage CP. The underlying concept of different ruminal solubilities further allows the CP fractions to be used in regression equations to estimate ruminally undegraded feed CP (RUP) values for a variety of feedstuffs, including forages [8,11].

There is a long history of research on forage conservation including N utilization in silage feeding [12]. However, there is a lack of targeted research on the conservation of herbage from temperate regions focusing on the relationship between conservation method and N balance in animals, as well as detailed descriptions of CP composition and protein value of the feed.

Therefore, the aim of this study was to determine the effect of three different conservation methods (i.e., ensiling, barn-drying, field-drying) of herbage on N balance and utilization in lactating dairy cows. We hypothesized that N utilization would be improved by feeding hay compared to silage due to its lower concentration of NPN. The differently conserved herbages were further characterized regarding CP composition and protein value, including CP fractionation and estimation of ruminal CP degradation and intestinal protein digestibility.

2. Materials and Methods

2.1. Preparation of Conserved Herbage

Herbage was cultivated at the experimental site Agroscope, Posieux, Switzerland (latitude: 46°46′ N, longitude: 07°06′ E; altitude: 650 m; 2016 average temperature: 9.2 °C; 2016 average precipitation: 1225 mm) in a sward mainly composed of *Lolium perenne* L., *Trifolium repens* L., and *Trifolium pretense* L. A 34 d regrowth was harvested as the fourth cut on 30 August 2016. One-third of the herbage was baled (0.8 × 0.7 × 1.3 m) without additives at a dry matter (DM) concentration of 56% after 24 h of wilting (silage, SI). A further third of the herbage, after 26 h on the field and at an average DM concentration of 68%, was put on ventilation (Hetroc dehumidifiers, Jona-Kempraten, Switzerland). In short, herbage was introduced into a hay box (basal area 6.2 × 9.9 m; volume 305 m^3) with a wooden grate. Ambient air was moderately heated (typically 5 to 8 °C above ambient temperature) with a heat pump and conducted through the material from below. The herbage was ventilated until a DM concentration of 88% was reached (barn-dried hay, BH). After 72 h of drying on the field, the rest was harvested at 86% DM and put on ventilation for one day (field-dried hay, FH). After drying, FH and BH were baled into square bales (0.8 × 0.7 × 2.2 m). During the harvesting period (30 August to 2 September 2016), the average values 2 m above ground for temperature, wind velocity, and sunshine duration were 19.9 °C, 1.5 m/s, and 535 min/d, respectively.

2.2. N Balance Trial

2.2.1. Trial Design and Animal Housing

Six multiparous Holstein cows were randomly assigned to three treatments (SI, BH, FH) in a replicated 3 × 3 Latin square arrangement. At the beginning of the trial, the cows were 270 ± 7 d in milk, had a body weight of 698 ± 65 kg and a milk yield of 23.5 ± 3.9 kg/d. Three consecutive 21 d experimental periods were conducted, each consisting of a 14 d adaptation and a 7 d data collection period. Cows were kept in a tie-stall barn with rubber mat flooring for the adaptation periods and transferred to metabolic cages during the data collection periods. Metabolic cages were equipped with rubber mat flooring and slatted floor in the anterior and posterior part of the cage, respectively.

SI, BH, or FH were fed to two cows each ad libitum during the adaptation periods. Feed residues were recorded daily, and feed intake was calculated. During the data collection periods, 0.95 of ad libitum feed intake was offered as a constant amount. Two cows receiving the same feed within one experimental period were randomly assigned, i.e., pairs were not kept together for the following experimental period. The cows received 300 g/d of a mineral mix containing 253, 92, 248, 147 g of ash, CP, neutral detergent fiber (NDF), and acid detergent fiber (ADF) per kg of DM, respectively, in two meals per day during the complete trial. The cows were milked twice a day at 7:00 and 16:00. All procedures were conducted in accordance with the Swiss guidelines for animal welfare and were approved (No. 2016_25_FR) by the Animal Care Committee of the Canton Fribourg, Fribourg, Switzerland.

2.2.2. Data Recording and Sample Collection

Body weight was determined during the adaptation periods after each milking when the cows left the milking parlor using a walk-through weight recording system with locking gates (Ga5010, Insentec B.V., PV Marknesse, The Netherlands). During the collection periods, each type of herbage was sampled daily to form two pooled samples per period for laboratory analyses (for SI, three pooled samples were formed in period 2 due to varying DM concentration). The samples were stored in plastic bags at −20 °C for SI and at room temperature for BH and FH. Feed residues were recorded daily. Milk yield was recorded at each milking, and milk samples were taken from each cow and handled for later analysis of gross constituents, urea, and N concentrations as described by Grosse Brinkhaus et al. [13]. Total feces were collected in a tub beneath the metabolic cage, and total urine was collected via urinals attached around the vulva via Velcro straps glued to the shaved skin. One part of the urine was acidified directly with 2.5 M sulfuric acid for later analysis of urinary N. Each morning, the total weights of feces and urine were measured. Feces were homogenized, and an aliquot of approximately 100 g was collected daily. For urine, 0.2% of the total daily amount was collected daily from the acidified collection vessels. In addition, aliquots of non-acidified urine were collected. Daily samples of both feces and urine were separately pooled per cow over each collection period and stored at −20 °C until further analysis. On d 1 and 7 of each collection period, at 7:00 before feeding, ruminal fluid was sampled via a stomach tube. At the same time points, blood was sampled from the jugular vein. Samples were prepared for later analysis of volatile fatty acids (VFA) and ammonia in ruminal fluid and urea in blood, as described by Grosse Brinkhaus et al. [13].

2.3. Laboratory Analyses

2.3.1. Silage Fermentation Quality Analysis

Silage pH was determined by inserting an electrode (No. 6.0202.110, Metrohm Schweiz AG, Zofingen, Switzerland) connected to an ion meter (pH/ionmeter 692, Metrohm Schweiz AG, Zofingen, Switzerland) into the filtered fluid extracted from 40 g samples shaken for 30 min with 400 mL of deionized water. The ammonia concentration of each extract was determined with an ammonia electrode (No. 6.0506.010, Metrohm Schweiz AG, Zofingen, Switzerland). Solutions of ~10 g silage, 90 mL deionized water, 2.5 mL Carrez I (18 g $K_4Fe(CN)_6 \times 3H_2O$ in 500 mL deionized water), 2.5 mL Carrez II (36 g $ZnSO_4 \times 7H_2O$ in 500 mL deionized water), and 5 mL internal standard solution were shaken (250 rpm) and extracted for 3 h. The concentrations of lactic, acetic, and butyric acid of the extracts were analyzed by high-performance liquid chromatography (HPLC; Summit, Thermo Fisher Scientific, Reinach, Switzerland) equipped with a nucleogel ION 300 OA 300 × 7.8 mm column (Macherey-Nagel, Düren, Germany) and a Shodex RI-101 refractive index detector (Shodex, Munich, Germany).

2.3.2. General Analyses

For chemical analysis, silage and feces samples were lyophilized (Christ, Osterode, Germany); all other feed samples were dried at 60 °C for 24 h. All samples were ground to pass a 1 mm screen (Brabender mill, Brabender, Duisburg, Germany). The DM and ash concentrations of feeds and feces were determined gravimetrically by oven-drying at 105 °C for 3 h and ashing at 550 °C until constant weight was attained (prepAsh, Precisa Instruments AG, Dietikon, Switzerland). Crude lipids concentrations were determined as petrol ether extract after an acidic hydrolysis in boiling HCl for 1 h (Method 5.1.1, VDLUFA [14]). NDF (Method 6.5.1 [14]; assayed with heat-stable amylase and without sodium sulfite, expressed without residual ash), ADF (Method 6.5.2 [14]; expressed without residual ash), and acid detergent lignin (Method 6.5.3 [14]) were analyzed using a Fibretherm analyzer (Gerhardt, Königswinter, Germany). The total N concentrations of feeds, feces, urine (acidified), and milk were analyzed using the Kjeldahl method (ISO 5983-1:2005) and—for the feed—multiplied by 6.25 to calculate the CP concentration. Water-soluble carbohydrates were determined as described by Hall et al. [15]. Milk samples were analyzed for fat, protein, and lactose concentrations using Fourier-transform infrared spectrometry (Milkoscan FT 6000, Foss, Hillerød, Denmark). Milk urea concentration was determined using the UreaFil test kit (MEA 549 EC Milk Urease, Eurochem, Moscow, Russia). Urinary (non-acidified) and serum urea concentrations were determined by enzymatic treatment with urease (EC 3.5.1.5) and glutamate dehydrogenase (EC 1.4.1.2) using a commercial test kit (No 147116, Greiner-Diagnostic, Langenthal, Switzerland). The ruminal VFA profile was analyzed by HPLC as described in Section 2.3.1. Ruminal ammonia was determined colorimetrically with a commercial test kit (Urea liquicolor, Human, Wiesbaden, Germany).

2.3.3. Ruminal Microbiota Quantitative PCR Analysis

DNA extraction was performed using QIAamp Fast DNA Stool Mini Kit (Qiagen, Hombrechtikon, Switzerland) following the manufacturer's instructions with slight modifications. Briefly, 2 mL of ruminal fluid were centrifuged at $6500 \times g$ for 30 min at 4 °C. The pellet was then resuspended in 2 mL of Inhibitex (provided with the mentioned kit) and heated at 90 °C for 5 min. The tubes were allowed to return to room temperature before 15 s vortexing and further centrifuged at $16,000 \times g$ for 1 min at room temperature. Afterwards, 200 µL of the supernatant were used for DNA extraction following the kit's procedure. DNA quantity was determined by spectrophotometry using a NanoDrop 1000 (Witec AG, Luzern, Switzerland). The quality of the extracted DNA was assessed by capillary electrophoresis using a Fragment Analyzer (Agilent technologies, Basel, Switzerland). The primers used in this study were previously described [13,16,17]. The primers were purchased in desalted quality (Microsynth, Balgach, Switzerland). Four micrograms of genomic DNA were used for amplification in the same conditions as previously described [13]. A reference sample was generated using a mixture of DNA derived from five different random ruminal fluids. The percentage of each considered strain in relation to total bacterial 16S ribosomal DNA (determined by amplification using GenBac primers) was calculated for the reference sample using the described formula [13]. For all the other samples, an induction fold was calculated relative to the abundance in the reference sample using a $\Delta\Delta Ct$ method with efficiency correction [18] and the EcoStudy software (Labgene, Châtel-Saint-Denis, Switzerland). The induction fold was then multiplied by the percentage calculated for the reference sample.

2.3.4. Feed Crude Protein Fractionation

Crude protein was categorized into five subfractions (i.e., A, B1, B2, B3, and C) based on the CNCPS [9]. For this purpose, TP, buffer-insoluble CP, neutral detergent-insoluble CP, and acid detergent-insoluble CP were specified according to standardizations of Licitra et al. [10] using Kjeldahl digestion to determine N (Method 4.1.1; VDLUFA [14]) All analyses were carried out in triplicate. In short, fraction A, which was NPN multiplied by 6.25, was calculated as CP minus TP precipitated with tungstic acid. Fraction B1 was TP soluble in borate-phosphate buffer. Fraction B2 was buffer-insoluble

CP minus neutral detergent-insoluble CP. Fraction B3 was neutral detergent-insoluble CP minus acid detergent-insoluble CP. Fraction C corresponded to acid detergent-insoluble CP.

2.3.5. Enzymatic In Vitro Estimation of RUP and RUP Intestinal Digestibility

Streptomyces griseus protease was used to simulate ruminal protein degradation and estimate RUP [19,20] following the forage-specific description of Edmunds et al. [8]. The samples were incubated for 1 h at 39 °C in borate-phosphate buffer before adding the protease solution (0.58 U/mL; Type XIV, ≥3.5 units/mg solid, P5147, Sigma-Aldrich, St. Louis, MO, USA) in an amount corresponding to 24 U/g TP determined using trichloroacetic acid as a precipitating agent [10]. After 24 h of incubation, the contents were filtered through a FibreBag (30 μm pore size, Gerhardt, Königswinter, Germany). In contrast to the procedure of Edmunds et al. [8], no vacuum was used for filtration, and rinsing of the FibreBags was replaced by washing in a beaker with fresh deionized water for 10 times [21]. FibreBags were freeze-dried, and the residues analyzed for N concentration using the Dumas combustion method (Method 4.1.2; VDLUFA [14]; rapid N cube, Elementar Analysensysteme, Hanau, Germany). Each sample was incubated in duplicate in two different runs, and RUP_{ENZ} (g/kg CP) was calculated as the amount of CP in the residue divided by incubated amount of CP, multiplied by 1000.

The residues from protease incubation were further used to estimate intestinal digestibility of RUP (IPD) [21,22]. The procedure was modified to account for the higher volume of residues from forage compared to concentrate by reducing the sample weight used for incubation and proportionately adjusting the enzyme dosage. In short, the residues were weighed into 50 mL centrifugation tubes in an amount including 7.5 mg N. After addition of 10 mL of a 0.1 N HCl solution (pH 1.9) containing 0.5 g/L of pepsin (P7012, Sigma-Aldrich, St. Louis, MO, USA), the tubes were incubated for 1 h in a shaking water bath at 38 °C. Subsequently, the solution was neutralized with 0.5 mL of 1 N NaOH, and 13.5 mL of phosphate buffer (pH 7.8; containing 1.5 g/L of pancreatin, P7545, Sigma-Aldrich, St. Louis, MO, USA) was added to each tube. After incubation for 24 h and vortexing every 8 h, 3 mL of trichloroacetic acid (1000 g/L) was pipetted into each tube to stop or minimize the enzymatic action and precipitate the undigested protein. The tubes were put on ice, and the contents were filtrated through filter paper (MN 640w, Macherey-Nagel, Düren, Germany). The residue on the filter paper was analyzed for insoluble N using the Kjeldahl procedure (Method 4.1.1, VDLUFA [14]). Pepsin–pancreatin incubation was carried out in triplicate. For calculation of IPD, N soluble in trichloroacetic acid was divided by N incubated in pepsin–pancreatin.

2.3.6. In Vitro Estimation of Utilizable Crude Protein at the Duodenum

A modified Hohenheim gas test was carried out to estimate utilizable CP at the duodenum (uCP). Based on the instructions of Menke and Steingass [23], modifications outlined by Steingaß and Südekum [24] and described in detail by Edmunds et al. [25] were applied. In short, 200–250 mg of feed was incubated in glass syringes with a ruminal fluid–buffer solution for 8 and 48 h. Ruminal fluid was obtained from two cannulated Holstein steers prior to morning feeding. The steers received a diet of grass hay (107 g CP and 5.40 MJ net energy for lactation (NEL) per kg DM) and concentrate feed (216 g CP and 7.6 MJ NEL per kg DM) in a ratio of 60:40 corresponding to their maintenance energy requirements. Each sample was incubated in duplicate for each time point within one run. Three runs were carried out, using ruminal fluid from different days. Additionally, two blanks containing only ruminal fluid–buffer solution were incubated for each time point within each run. After incubation, syringes where put on ice in order to stop the fermentation, and the quantity (mg) of ammonia-N was measured in the samples (ammonia-N_{sample}) and blanks (ammonia-N_{blank}) using automated distillation (Vapodest 50 s carousel, Gerhardt, Königswinter, Germany). For both 8 and 48 h, uCP (g/kg DM) was calculated as follows:

$$uCP = ((N_{sample} - (ammonia\text{-}N_{sample} - ammonia\text{-}N_{blank}))/weight_{sample}) \times 6.25 \times 1000, \qquad (1)$$

where N_{sample} is total N added by sample (mg), weight$_{sample}$ is the amount of sample incubated expressed as mg DM, and other variables are described above. Linear regression of uCP values at 8 and 48 h to the natural logarithm (ln) of time allowed for the calculation of effective uCP for an assumed ruminal passage rate (K_p) of 0.05/h through calculating the function value of ln (20).

2.4. Calculations and Statistical Analysis

The concentrations of NEL and absorbable protein in the small intestine (APD) in the conserved herbages were estimated according to Swiss nutrient recommendations for ruminants (Agroscope [26]). First, organic matter (OM) digestibility (%) was calculated on the basis of regression equations for balanced mixed swards including mainly ryegrass [26]:

$$\text{OM digestibility of silage} = 16.9 + 0.0864\ CP + 0.3815\ ADF - 0.000125\ CP^2 - 0.000755\ ADF^2, \quad (2)$$

$$\text{OM digestibility of hay} = 27.3 + 0.0924\ CP + 0.2846\ ADF - 0.000162\ CP^2 - 0.000581\ ADF^2, \quad (3)$$

where CP and ADF are in g/kg of OM. In the results section, the calculated OM digestibility was expressed as a coefficient.

To further estimate NEL concentrations (MJ/kg DM), metabolizable energy (ME)—estimated from the calculated OM digestibility—and gross energy (GE)—estimated from OM and CP concentrations—were used [26]:

$$\text{NEL} = (0.463 + 0.24\ \text{ME/GE}) \times \text{ME} \times 0.9752, \quad (4)$$

The APD (g/kg DM), when ruminally fermentable energy (APDE) or N (APDN) limits microbial protein synthesis in the rumen, was calculated as follows [26]:

$$\text{APDE} = 0.093 \times \text{FOM} + CP \times (1.11 \times (1 - \text{deCP}/100)) \times \text{dAAF}/100, \quad (5)$$

$$\text{APDN} = CP \times (\text{deCP}/100 - 0.10) \times 0.64 + CP \times (1.11 \times (1 - \text{deCP}/100)) \times \text{dAAF}/100, \quad (6)$$

where CP is given in g/kg DM, FOM is fermentable OM (g/kg DM), deCP is degradability of CP (%), and dAAF is digestibility of amino acids (AA) in the feed (%). The values of FOM, deCP, and dAAF were calculated according to Agroscope [26].

Assuming a K_p of 0.05/h, RUP was estimated from chemical CP fractionation (RUP$_{CHE}$; g/kg CP) on the basis of the equation of Kirchhof [11]:

$$\text{RUP}_{CHE} = 321.9023 + (0.1676 \times \text{PADF}) + (-0.0022 \times (CP \times (A + B1))) + (0.0001 \times (CP \times C^2)), \quad (7)$$

where PADF (g/kg DM) refers to ADF estimated from the residue after boiling in acid detergent solution according to Licitra et al. [10], CP is in g/kg DM, and CP fractions are in g/kg CP.

The potential prececal CP digestibility (fraction of CP) was calculated from CP concentration (g/kg DM), RUP$_{ENZ}$ (g/kg CP), and IPD (fraction of RUP) estimated in vitro as follows:

$$\text{Potential prececal CP digestibility} = (CP \times (1000 - \text{RUP}_{ENZ})/1000 + (CP \times \text{RUP}_{ENZ}/1000 \times \text{IPD}))/CP, \quad (8)$$

The apparent total tract digestibility of OM was calculated from the daily amounts of OM in feed and feces and then used to calculate the intake of digestible OM. Nitrogen balance was calculated as N intake minus N excretion via milk, urine, and feces and expressed as g/d. Balances of uCP and APDE were calculated as dry matter intake (DMI) × feed concentration of uCP or APDE, respectively, minus the requirements estimated from the German feed evaluation system (GfE [27]) and from Agroscope [26] for uCP and APDE, respectively. Energy-corrected milk yield (ECM) was calculated on a 4.0% fat, 3.2% protein, and 4.8% lactose basis [26].

Statistical analysis was done using the SAS 9.4 software (SAS Institute Inc., Cary, NC, USA). Data on feed intake and digestibility, N intake and excretion, milk yield and composition, ruminal VFA and ammonia concentrations, serum urea concentrations and ruminal microbiota were analyzed with PROC MIXED of SAS with conservation method and experimental period as fixed effects and cow as random effect. The results are expressed as least-squares means, and the differences were tested with Tukey's test. Significance was defined at $p < 0.05$, and tendencies were declared for $p = 0.05$ to $p < 0.10$.

3. Results

3.1. Characteristics of Conserved Herbage

Silage displayed a pH of 5.5, and the concentrations of lactic, acetic, and butyric acid were 27, 5 and 1 g/kg DM, respectively. The results on the chemical composition of the forages are given in Table 1. The chemical composition was similar between types of forage, with NDF revealing the highest variation. The concentrations of CP and NEL (Table 1) as well as the calculated OM digestibility (Table 1) were slightly higher in SI compared to both types of hay. Silage also displayed the highest APDN but the lowest APDE concentrations (Table 1). The concentrations of APDN were higher than those of APDE for all types of forage (Table 1).

Table 1. Dry matter (DM) concentration and chemical composition as well as calculated values [26] of organic matter digestibility, concentrations of net energy for lactation, and absorbable protein at the duodenum of silage (SI; n = 7), barn-dried hay (BH; n = 6), and field-dried hay (FH; n = 6). Values are reported as means ± standard deviation of pooled samples of the same type of forage.

	SI	BH	FH
DM (g/kg)	554 ± 82.8	877 ± 9.6	868 ± 10.2
Ash (g/kg DM)	119 ± 7.7	114 ± 2.0	109 ± 2.0
Crude protein (g/kg DM)	207 ± 6.6	187 ± 3.8	176 ± 3.4
Crude lipids (g/kg DM)	45.7 ± 5.40	42.5 ± 3.08	35.0 ± 2.10
Neutral detergent fiber [1,2] (g/kg DM)	406 ± 13.4	438 ± 6.4	482 ± 22.0
Acid detergent fiber [2] (g/kg DM)	260 ± 11.6	268 ± 6.4	283 ± 4.1
Acid detergent lignin (g/kg DM)	21.8 ± 2.86	21.5 ± 2.43	22.8 ± 3.06
Water-soluble carbohydrates (g/kg DM)	67.5 ± 19.68	80.5 ± 3.94	72.1 ± 2.14
Organic matter digestibility	0.743 ± 0.0084	0.703 ± 0.0028	0.687 ± 0.0067
Net energy for lactation (MJ/kg DM)	5.96 ± 0.098	5.53 ± 0.052	5.38 ± 0.075
APDE (g/kg DM)	89.2 ± 4.39	97.0 ± 0.93	94.2 ± 1.03
APDN (g/kg DM)	130 ± 3.9	120 ± 2.3	112 ± 2.1

APDE/APDN, absorbable protein at the duodenum when ruminally fermentable energy (APDE) or N (APDN) limits microbial protein synthesis in the rumen. [1] Assayed with heat-stable amylase. [2] Expressed without residual ash.

Crude protein fractions B1 and C were similar for all types of forages, whereas the other fractions showed some variation (Table 2). Specifically, CP fraction A was almost 200 g/kg CP higher in SI compared to BH and FH. Crude protein fraction B3 was lowest in SI, highest in FH, and intermediate in BH. The estimated concentrations of uCP were on average 157 g/kg DM (Table 2). The estimation from both CP fractions and the enzymatic in vitro method resulted in a similar pattern of RUP values, with the lowest values for SI, the highest values for FH, and BH being intermediate (Table 2). The estimated IPD was below 0.50 for SI, BH, and FH (Table 2).

Table 2. Crude protein (CP) fractions and chemically and in vitro estimated characteristics of the protein value of SI (n = 7), BH (n = 6), and FH (n = 6). Values are reported as means ± standard deviation of pooled samples of the same type of forage.

	SI	BH	FH
Crude protein fractions (g/kg CP)			
A	448 ± 53.4	260 ± 28.7	256 ± 13.2
B1	32.1 ± 14.10	54.8 ± 32.86	29.4 ± 14.23
B2	333 ± 35.2	421 ± 16.2	370 ± 16.0
B3	124 ± 33.3	200 ± 19.6	278 ± 13.5
C	63.4 ± 6.82	64.3 ± 12.80	66.7 ± 10.31
Protein value characteristics			
uCP (g/kg DM)	144 ± 15.7	160 ± 13.0	169 ± 17.2
RUP$_{CHE}$ (g/kg CP)	238 ± 37.9	322 ± 32.6	344 ± 24.5
RUP$_{ENZ}$ (g/kg CP)	316 ± 32.8	363 ± 27.7	393 ± 27.7
IPD	0.47 ± 0.062	0.49 ± 0.077	0.43 ± 0.091
Potential preceal digestibility of CP [1]	0.83 ± 0.018	0.81 ± 0.018	0.78 ± 0.031

A, B1, B2, B3, C, CP fractions estimated according to Licitra et al. [10]; uCP, utilizable CP at the duodenum estimated from in vitro incubation [25]; RUP, ruminally undegraded feed CP estimated from chemical CP fractionation (RUP$_{CHE}$; [11]) or in vitro protease incubation (RUP$_{ENZ}$; [8]); IPD, intestinal digestibility of RUP estimated from pepsin–pancreatin incubation [22]. [1] Calculated from CP concentration, RUP$_{ENZ}$, and IPD.

3.2. Feed Intake and Digestibility

Dry matter intake during the collection periods tended to be lower ($p = 0.05$) for SI compared to BH, while the intake of FH did not differ from the other treatments (Table 3). The apparent total tract digestibility of OM was lower for FH compared to SI ($p = 0.01$; Table 3). The apparent digestibility of NDF was not affected by the conservation method, but the apparent digestibility of ADF was or tended to be higher for SI compared to FH ($p = 0.02$) and BH ($p = 0.09$). The intake of digestible OM tended to be higher for BH compared to SI ($p = 0.08$) and FH ($p = 0.09$).

Table 3. DM intake, apparent total tract digestibility of organic matter (OM), neutral detergent fiber (NDF), acid detergent fiber (ADF), and intake of apparently digestible OM in cows fed SI, BH, or FH.

	SI	BH	FH	SEM	*p*-Value
Feed intake [1,2] (kg DM/d)	17.3	19.2	17.9	0.71	0.05
Apparent OM digestibility [2]	0.743 [a]	0.730 [ab]	0.712 [b]	0.0067	0.01
Apparent NDF digestibility [2]	0.750	0.740	0.737	0.0078	0.26
Apparent ADF digestibility [2]	0.771 [a]	0.748 [ab]	0.738 [b]	0.0091	0.02
Intake of digestible OM (kg/d) [2,3]	11.3	12.4	11.3	0.45	0.06

Values with different superscripts within a row differ ($p < 0.05$). SEM, Standard error of the mean. [1] Feed intake during the collection periods, i.e., feed offering was adjusted to 0.95 of ad libitum feed intake during the adaptation periods. [2] Calculations of feed intake and digestibility include the contribution from mineral feed. [3] Calculated from feed intake, proportion of OM in DM, and apparent OM digestibility.

3.3. N Intake, Digestibility, and Excretion in Milk, Urine, and Feces

Urine volume was not statistically different between treatments and was, on average, 46.3 L/d (Table 4). Table 4 shows the results for N intake, digestibility, and excretion. The daily N intake of cows fed SI was not different compared to the intake of cows fed BH or FH, but cows fed BH had higher ($p = 0.02$) N intake than cows fed FH. Feeding FH compared to SI ($p = 0.001$) or BH ($p = 0.01$) resulted in lower apparent total tract digestibility of N. The excretion of N in milk tended to be higher for cows fed BH compared to those fed FH ($p = 0.08$) and SI ($p = 0.07$). Fecal N excretion was 177 g/d on average and similar for all forage types. Urinary N excretion (g/d) tended to be higher ($p = 0.06$) when cows were fed SI compared to FH but was not different for BH compared to the other conservation methods. Urinary N excretion expressed as proportion of N intake did not differ between treatments. The N balance was negative (−27.2 g/d, on average) for all cows and not affected by the conservation method. Fecal N expressed as a proportion of N intake was higher for cows fed FH compared to cows fed SI ($p = 0.001$) and BH ($p = 0.01$). The proportion of milk N of total N intake (N use efficiency, NUE) did not differ between treatments and was, on average, 21%. The urine of cows fed SI ($p = 0.01$) and BH ($p = 0.04$) had higher concentrations of urea compared to the urine of cows fed FH. Also, the amount (g/d) of urinary N excretion in the form of urea (UUN) was higher for cows fed SI ($p = 0.01$) and tended to be higher ($p = 0.05$) for cows fed BH compared to FH. The daily excretion of urinary non-urea N (UNUN) was not affected by the conservation method and, on average, comprised 0.19 of total urinary N excretion.

Table 4. N intake and apparent total tract digestibility and excretion of N in milk, urine, and feces, as well as urine volume for cows fed SI, BH, or FH.

	SI	BH	FH	SEM	*p*-Value
N intake (g/d)	560 [ab]	581 [a]	509 [b]	21.9	0.03
Apparent N digestibility [1]	0.701 [a]	0.685 [a]	0.647 [b]	0.0082	0.001
Urine (L/d)	48.1	47.5	43.3	1.97	0.21
N excretion (g/d)					
Milk N	112	123	113	6.0	0.05
Urinary N	307	295	251	15.9	0.06
Fecal N	168	183	180	8.6	0.34
Total N	587	600	543	19.8	0.13
N balance	−27.4	−19.4	−34.8	19.4	0.85
N excretion (% of N intake) [2]					
Fecal N	29.9 [a]	31.5 [a]	35.3 [b]	0.83	0.001
Urinary N	55.0	51.8	49.2	3.73	0.55
Milk N	20.2	21.2	22.2	0.87	0.11
Fractionation of urinary N					
Urinary urea (mmol/L)	188 [a]	181 [a]	163 [b]	6.4	0.01
Urinary urea N (g/d)	253 [a]	241 [ab]	198 [b]	11.7	0.01
Urinary urea N/Urinary N	0.823	0.821	0.791	0.021	0.51
Urinary non-urea N (g/d)	54.3	53.7	53.1	7.36	0.99
Urinary non-ureaN/Urinary N	0.177	0.180	0.209	0.0206	0.51

Values with different superscripts within a row differ ($p < 0.05$). SEM, Standard error of the mean. [1] Calculation of digestibility includes the contribution from mineral feed. [2] Negative N balance results in total excretion amounting to >100%.

3.4. Milk Yield and Composition

The results for milk yield and composition are shown in Table 5. The type of forage had no significant effect on milk fat and protein percentages. There was a tendency for milk yield ($p = 0.09$) and ECM ($p = 0.09$) to be higher for BH compared to FH. Milk protein yield tended to be higher when the cows were fed BH compared to SI ($p = 0.07$) and FH ($p = 0.08$). Milk urea concentration was higher for cows fed SI ($p < 0.001$) or BH ($p = 0.003$) compared to cows fed FH.

Table 5. Milk yield and composition for cows fed SI, BH, or FH.

	SI	BH	FH	SEM	*p*-Value
Milk yield (kg/d)	19.3	20.6	19.0	1.38	0.09
ECM (kg/d)	22.1	23.5	21.8	1.80	0.09
Milk components (%)					
Fat	4.98	4.88	4.93	0.147	0.35
Protein	3.73	3.83	3.82	0.099	0.18
Lactose	4.70	4.71	4.68	0.073	0.74
Milk urea (mg/kg)	370[a]	351 [a]	306 [b]	14.9	<0.001
Milk component yield (kg/d)					
Fat	0.968	1.012	0.947	0.0911	0.11
Protein	0.717	0.785	0.720	0.0380	0.05
Lactose	0.911	0.977	0.899	0.0786	0.12

Values with different superscripts within a row differ ($p < 0.05$). ECM, energy-corrected milk yield [26]. SEM, Standard error of the mean.

3.5. Ruminal Fluid Ammonia and Volatile Fatty Acids and Serum Urea

Ammonia concentration was higher ($p = 0.04$) in the ruminal fluid of cows fed BH compared to cows fed FH, whereas it was not different for SI compared to BH and FH (Table 6). The total concentration of VFA in the ruminal fluid as well as the proportions of individual VFA did not differ between treatments (Table 6). Serum urea concentration was higher ($p = 0.001$) in cows fed BH and SI compared to those fed FH (Table 6).

Table 6. Concentrations of ruminal ammonia and volatile fatty acids (VFA) as well as serum urea in cows fed SI, BH, or FH.

	SI	BH	FH	SEM	*p*-Value
Ruminal ammonia (mmol/L)	7.38 [ab]	8.15 [a]	6.98 [b]	0.423	0.04
Total VFA (mmol/L)	82.5	87.0	86.1	4.82	0.67
VFA molar proportion (%)					
Acetate	70.8	70.4	71.1	0.23	0.19
Propionate	14.7	14.8	14.9	0.16	0.66
n-Butyrate	10.7	10.8	10.4	0.18	0.19
Isobutyrate	1.26	1.30	1.19	0.052	0.33
n-Valerate	1.10	1.11	1.04	0.034	0.25
Isovalerate	1.35	1.53	1.36	0.066	0.14
Acetate:propionate ratio	4.82	4.77	4.76	0.056	0.77
Serum urea (mmol/L)	7.23 [a]	7.22 [a]	6.45 [b]	0.202	<0.001

Values with different superscripts within a row differ ($p < 0.05$). SEM, Standard error of the mean.

3.6. Ruminal Microbiota Quantification

Feeding differently conserved herbages did not affect the relative abundances of *Lactobacillus* spp. and *Fibrobacter succinogenes* but influenced those of the other examined bacterial species (Table 7). When cows were fed SI compared to FH, *Prevotella* spp. displayed higher ($p = 0.02$) relative abundances, while the levels of *Butyrivibrio fibrisolvens* were lower ($p = 0.01$). *B. fibrisolvens* relative abundance also tended ($p = 0.07$) to be lower in the ruminal fluid of cows fed BH compared to FH. The abundances of the *Ruminococcus* species *albus* ($p = 0.03$) and *flavefaciens* ($p = 0.04$) were lower when feeding BH compared to FH. For *R. albus*, a lower relative abundance was also observed in the ruminal fluid of cows fed BH compared to SI ($p = 0.01$).

Table 7. Relative abundance (% of total 16S DNA) of ruminal bacteria species in the ruminal fluid of cows fed SI, BH, or FH.

	SI	BH	FH	SEM	*p*-Value
Lactobacillus spp.	0.115	0.102	0.115	0.0048	0.12
Prevotella spp.	48.5 [a]	45.4 [ab]	41.3 [b]	1.54	0.03
Butyrivibrio fibrisolvens	0.0333 [a]	0.0378 [ab]	0.0476 [b]	0.00281	0.01
Fibrobacter succinogenes	6.00	5.98	6.67	0.470	0.51
Ruminococcus albus	5.81 [a]	4.02 [b]	5.46 [a]	0.551	0.01
Ruminococcus flavefaciens	13.9 [ab]	12.4 [a]	16.6 [b]	1.29	0.0478

Values with different superscripts within a row differ ($p < 0.05$). SEM, Standard error of the mean.

4. Discussion

The feed characteristics of the differently conserved herbages reflected typical effects of the conservation method, such as lower and higher concentrations of DM and CP fraction A, respectively, in silages compared to hay. A trend towards lower CP and higher fiber concentrations from SI over BH to FH could be related to longer wilting time, increased mechanical processing, and an associated loss of leaf material. Generally, the production of barn-dried hay can result in considerably lower DM losses from cutting to feeding compared to field-dried hay and, in some cases, also to silage [28]. The duration until inhibition of respiration either by anaerobic conditions in silage or by low moisture in hay has a large impact on forage quality [28,29]. Consequently, NEL concentrations were the highest for SI and the lowest for FH, which underwent the longest time until stable DM conditions were reached.

The silage had a relatively high DM concentration and, therefore, the fermentation process was limited, as reflected in the low concentration of lactic acid. However, silage fermentation quality was "very good" when assessed with the scheme of the German Agricultural Society (DLG [30]) based on the concentrations of acetic acid and butyric acid and the pH value. Fermentation quality is linked to DM concentration in grass silages, which is why higher DM concentrations can increase feed intake [31]. A considerably higher feed intake for hay compared to silage has been reported [32,33]. However, the effect is dependent on a variety of characteristics often related to silage quality [34], not clearly demonstrated by literature data [2] and, in the current study, was visible only for BH but not for FH. The intake by sheep was higher for barn-dried hay compared to field-dried hay, possibly due to higher OM digestibility [35]. In the current study, apparent total tract OM digestibility was not significantly different between FH and BH, but it was higher for SI compared to FH. This could be related to lower concentration (NDF) and higher digestibility (ADF) of fiber in SI. Higher CP (or N) digestibility and, specifically, degradation in the rumen [36] may also have contributed to higher OM digestibility in SI, but quantitative aspects of ruminal OM or CP degradation were not investigated here. However, ruminal VFA concentrations were analyzed. Friggens et al. [37] discussed considerable differences in the molar proportions of VFA in the ruminal fluid when feeding silage versus hay. In other studies, moderate effects on single VFA were observed [38,39]. In contrast, neither total concentrations nor molar proportions of VFA were affected by the conservation method in the current study. The lack of effect may be due to the silage being relatively dry and restrictedly fermented and thus yielding lower propionate proportions of VFA compared to extensively fermented silages [39]. The proportions of acetate and propionate were higher and lower, respectively, than reported by other authors when comparing hay and silage feeding [38,39], presumably due to an at least moderate supplementation of concentrate in the latter studies.

In addition to VFA, the relative abundances of ruminal bacteria were evaluated via quantitative PCR. *Prevotella* spp. represented the majority of bacteria, consistent with earlier observations [40]. The higher abundance in the ruminal fluid from cows fed SI compared to FH could be related to better accessibility of feed protein, as *Prevotella* are known for proteolytic and peptidolytic activity [41]. *B. fibrisolvens* also displays proteolytic activity [41] and was elevated through feeding FH compared to SI, but its overall relative abundances were low. Carbohydrate-degrading bacteria *F. succinognenes*,

R. albus, and *R. flavefaciens* showed together a high relative abundance, presumably reflecting the forage-only diet. The conservation method affected *R. albus* and *R. flavefaciens* relative abundances, but the effect was not consistent.

The focus of our study was on N balance and utilization. Dry matter intake of feed from all conservation methods was high, given the cows were in late lactation. Combined with the high dietary CP concentrations, this led to high intake of N. Further, intake of uCP calculated from the concentration of uCP estimated in vitro and feed intake exceeded uCP requirements [27] (Figure 1). Similarly, APD intake was in excess with respect to the requirements [26], except for two cows in two periods (Figure 1).

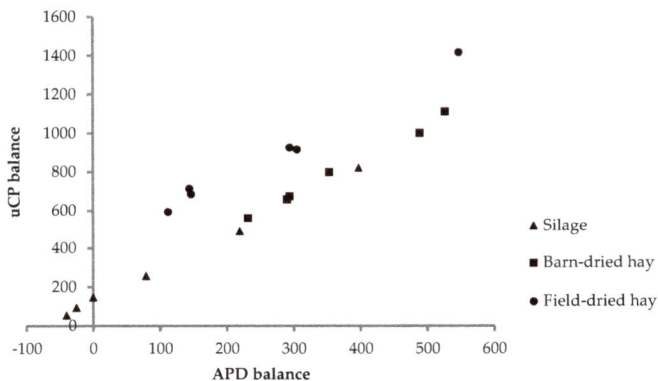

Figure 1. Balances (estimated intake minus requirements; g/d) of utilizable crude protein at the duodenum (uCP) [27] and absorbable protein at the duodenum (APD; calculated from APD when ruminally fermentable energy limits microbial protein synthesis in the rumen, i.e., APDE [26]). Each data point represents one cow in one experimental period.

As a consequence of excess dietary N intake, urinary N excretion amounted to around 0.5 of N intake. Urine as the main route of surplus N excretion has been observed for various diet compositions (e.g., [42–44]). Regulatory N excretion via urine is in the form of urea [45], which could also be observed in the current experiment, where feeding FH resulted in both the lowest N intake and the lowest UUN excretion. In contrast, UNUN excretion was similar for all treatments and as such not affected by different N intakes. The observed UNUN excretion was also very close to the value of 51.9 g/d estimated by Spek et al. [46]. Moreover, UNUN was around 3 g/kg DMI and thus consistent with the values reported in a literature review by Pfeffer et al. [45], who concluded that UNUN excretion mostly is <4 g/kg DMI.

Similar to UNUN, fecal N excretion did not differ between treatments and was approximately 10 g/kg DMI. This value is in line with collated literature data and can be viewed as obligatory and not related to the regulation of N in the body pool [45]. However, fecal N excretion expressed as proportion of N intake decreases with higher N intake [47]. Thus, a higher percentage of fecal N excretion related to N intake for FH was most likely an artifact of slight differences in feed intake and CP concentration of the herbages, resulting in lower daily N intake for FH. In contrast to fecal N excretion, the proportion of UNUN in urinary N was not significantly affected by the treatment, which is not consistent with the concept of UNUN seen as obligatory excretion. The fact that less N had to be disposed of when FH was fed was visible not only in UUN excretion but also in lower urea concentrations in serum and milk of cows fed FH.

Even though differences in N intake certainly explain a significant share of the observed effects on urea concentrations, feed protein characteristics may also play a role. Field-dried hay displayed the lowest apparent total tract digestibility of N. This is in line with the lower potential of prececal

CP digestibility indicated by the enzymatic in vitro method. Moreover, RUP values estimated from CP fractionation and in vitro incubation in *S. griseus* protease solution indicated lower ruminal CP degradability for FH, which was also reflected in lower ruminal ammonia concentration for FH compared to BH. However, SI displayed higher concentrations of CP, which in addition contained a higher proportion of NPN. This should theoretically have led to higher ruminal ammonia concentrations for SI compared to FH, which was not the case. A possible explanation could be the fact that the ruminal fluid was collected before the morning feeding. At this time point, ammonia rapidly released from dietary NPN can already be absorbed. Moreover, significant amounts of soluble N fractions from silage, including non-ammonia NPN, may escape from the rumen [48]. In contrast, protein degradation in BH and FH will have proceeded more steadily.

Crude protein fractionation revealed the highest proportion of fraction B3 in FH. In a study by Edmunds et al. [8], 60% of the variation in RUP in silages and dried forages could be assigned to differences in CP fraction B3. In sheep, Verbič et al. [36] found a lower ruminal CP degradability of hay compared to differently prepared silages from the same parent material. However, it cannot be determined if the observed results indicating lower N turnover and clearance for FH were due to lower N intake, lower ruminal CP degradability, or a combined effect. While the effect of increased dietary RUP concentration is debated, reducing N supply is a commonly recommended measure to reduce N excretion and increase NUE [47], also in grass-based diets [49,50].

Feeding SI and BH resulted in similar N intake. The observed pattern of CP fractions and in vitro and chemical RUP estimation indicated that BH delivered higher amounts of RUP to the animals, which could have led to the observed tendencies for higher milk N excretion and milk protein yield in cows fed BH compared to SI. However, there was also a tendency towards a higher intake of feed and particularly digestible OM for BH compared to SI. On the other hand, neither partitioning of N excretion nor urea concentrations in milk and serum differed between feeding BH and SI, contradictory to our hypothesis that N utilization would be improved by feeding hay compared to silage. True protein in SI still contributed >500 g/kg CP. Much lower TP concentrations can be reached as a result of protein breakdown even in well-fermented silages [51]. The silage produced in this experiment was relatively dry due to constant dry weather conditions during the wilting period on the field. Possibly, stronger effects of ensiling compared to drying of herbage could have been expected if silage with lower DM concentrations had been produced. In particular, concentrations of TP, RUP, and uCP may be lower in silages with lower DM concentrations [52]. Furthermore, DM concentrations in grass-clover silages are positively correlated to the duodenal flow of microbial CP [53]. However, even if RUP supply was actually different between SI and BH, excess supply of feed CP by both treatments may have prevented possible positive effects of an increased dietary RUP concentration on NUE [54].

The enzymatic estimation revealed similar but low IPD values for herbage from all three conservation methods. This indicates that a large proportion of RUP consisted of fiber-bound N and, hence, was not accessible for enzymatic digestion in the small intestine. However, IPD was lower than the values for grass products reported in the literature [55,56], but it has to be noted that methods differed. Edmunds et al. [57] demonstrated that the AA pattern of forage protein is altered during ruminal incubation but does not widely differ between RUP from differently conserved forages. For the current study, this would imply that only total supply and not quality in terms of intestinal digestibility and AA pattern of RUP differed between forages.

Overall, NUE was low, as N excreted in milk was only 20–22% of N intake. These values correspond well to the efficiency of N utilization observed for the lower quartile in collated data of Calsamiglia et al. [58]. Interestingly, the CP concentration in forage that Calsamiglia et al. [58] estimated for this quartile was almost equal to the CP concentration of the conserved forages in the current experiment. For diets mainly based on grass silage, NUE estimated from collated feeding trial data was 27.7% [59]. However, the reported minimum and maximum NUE values were as low as 16.0 and as high as 40.2%, respectively [59]. Reports of NUE in dairy cows receiving only conserved forage are scarce. Shingfield et al. [39] observed slightly higher NUE for hay compared to differently treated silages prepared from the same mixed swards (timothy and meadow fescue). However, concentrate supplementation was part of the experiment, and the level of NUE was around 30%. A similar mean value was demonstrated for cows fed grass-clover silage supplemented with concentrate [60]. Low NUE of around 20–25% were also reported for cows grazing ryegrass pasture with only moderate concentrate supplementation [61].

The utilization of N seems particularly low, given the fact that adequate or surplus supply of APD and uCP was accompanied by a negative N balance. Moreover, N intake of cows largely exceeded the requirements to maintain a stable N balance calculated by Pfeffer et al. [45]. A negative N balance indicates a mobilization of body protein. This can occur during non-sufficient dietary supply of N, when AA from the skeletal muscle protein are used for milk protein synthesis [62]. However, in the current experiment, a shortage of dietary N supply was precisely not likely, and thus AA from degraded body protein would not have been essential for milk protein synthesis. Instead, it is more likely that AA from skeletal muscle protein were used for energy supply [62], and the amino group of AA was disposed of as urea and excreted via urine. In fact, except for one cow, all cows lost body weight over the course of the complete trial (body weight change from −49.9 to +1.9 kg, average −23.6 kg). Milk yields were moderate, but high milk protein and fat concentrations elevated ECM. The requirements of NEL [27] were not met by the actual intake for five of the six cows, and the mean estimated NEL balance was −9.2 MJ/d. Hence, the assumptions of Pfeffer et al. [45] regarding N supply to maintain a stable N balance were not met in this study. Negative energy balance is of major significance during early lactation, when substantial amounts of body protein can be mobilized along with body fat despite sufficient dietary CP supply [63]. However, the proportion of mobilized body protein in total mobilized tissue decreases fast after parturition, and protein balance can become positive after four weeks of lactation [64]. In contrast, the cows in the current experiment were in late lactation, where energy supply under most feeding regimes is not limited.

Balancing dietary energy and protein supply to maximize N utilization is primarily discussed concerning ruminal metabolism [65]. Energy supply matching N supply may lead to the efficient use of N for microbial growth and help in capturing rapidly released ammonia, e.g., in silages. However, the current results should also be seen in the light of adequate postabsorptive energy supply, which may improve AA uptake in the mammary gland independently from protein supply [66]. In this regard, Tamminga [67] discussed postabsorptive N losses due to an imbalance between energy and AA availability at the tissue level. This also has a practical implication for herbage-dominated feeding systems without supplementary concentrate. These systems can result in a "high metabolic load in high-yielding dairy cows during early lactation" [68]. Although the cows in the current study were not in early lactation, and milk yields were moderate, the loss of body weight and a negative NEL balance point to the fact that high metabolic loads may have occurred nonetheless.

When feeding only forage, balancing the supply of energy and CP is a challenge. Harvest and conservation of herbage are weather-dependent and thus offer limited opportunities to modify both energy and CP concentrations to the desired level. From a study with grass silages, Dewhurst et al. [69] clearly concluded that in order to maximize the utilization of grass silage N, crops with higher energy and lower CP concentrations are needed. The results from the current experiments underline this conclusion and further indicate that it can be extended to herbage conserved as hay. Similar suggestions have been made with regard to pasture systems, where high N concentrations in ryegrass and clover result in high N losses [49]. Energy may become first limiting, and N be used less efficiently when cows are fed pasture without supplementation [70].

5. Conclusions

Although the cows in this study were in late lactation, feeding only forage derived from herbage resulted in negative N and energy balances regardless of the method of conservation. Contradictory to our hypothesis, the utilization of feed N for milk N was not different between cows fed SI, BH, and FH. From chemical and in vitro estimations, it could be concluded that the conservation method had considerable effects on CP composition and protein value of the forages. These differences were not or only moderately reflected in the animals' responses, which can be explained by the fact that N supply exceeded the requirements for all three treatments. Lower urea concentrations in serum, milk, and urine when FH was fed were likely due to the lower N intake observed for FH. The effects of the different conservation methods will be presumably more pronounced when (i) silage exhibits lower DM concentration, (ii) the supply of total CP, APD, and uCP is not in excess, and (iii) the energy supply is not limited. This has implications for future research on comparing forage conservation methods, e.g., silage DM concentration and energy supply have to be considered in new study designs. In addition, these aspects should be addressed in practical feeding situations where dairy cows are fed solely on herbage.

Author Contributions: Conceptualization, F.D.-M., K.-H.S., and U.W.; methodology, C.B., P.S., F.D.-M., K.-H.S., and U.W.; validation, C.B., P.S., F.D.-M., K.-H.S., and U.W.; formal analysis, C.B., P.S., F.D.-M., and U.W.; investigation, C.B., P.S., and U.W.; data curation, C.B., P.S., and U.W.; writing—original draft preparation, C.B.; writing—review and editing, C.B., P.S., F.D.-M., K.-H.S., and U.W.; visualization, C.B.; supervision, F.D.-M. and K.-H.S.; project administration, F.D.-M. and U.W.; funding acquisition, F.D.-M. and U.W.

Funding: This research was funded by Arbeitsgemeinschaft zur Förderung des Futterbaues AGFF, Zürich, Switzerland.

Acknowledgments: The authors thank the animal husbandry and laboratory staff at Agroscope Posieux and Institute of Animal Science, Bonn, and J.-L. Oberson for his advice in conducting the N balance trial. D. Latzke and A. Saßenbach are gratefully acknowledged for carrying out the modified Hohenheim gas test and the enzymatic in vitro method, respectively.

Conflicts of Interest: The authors declare no conflict of interest. The funders had no role in the design of the study; in the collection, analyses, or interpretation of data; in the writing of the manuscript, or in the decision to publish the results.

Abbreviations

A	crude protein fraction according to the Cornell Net Carbohydrate and Protein System
AA	amino acid(s)
ADF	acid detergent fiber
APD	absorbable protein in the small intestine
APDE	absorbable protein in the small intestine when ruminally fermentable energy limits microbial protein synthesis in the rumen
APDN	absorbable protein in the small intestine when N limits microbial protein synthesis in the rumen
B1	crude protein fraction according to the Cornell Net Carbohydrate and Protein System
B2	crude protein fraction according to the Cornell Net Carbohydrate and Protein System
B3	crude protein fraction according to the Cornell Net Carbohydrate and Protein System
BH	barn-dried hay
C	crude protein fraction according to the Cornell Net Carbohydrate and Protein System
CNCPS	Cornell Net Carbohydrate and Protein System
CP	crude protein
DM	dry matter
dAAF	digestibility of amino acids in the feed
deCP	degradability of crude protein
DMI	dry matter intake
ECM	energy-corrected milk yield
FH	field-dried hay
FOM	fermentable organic matter
GE	gross energy
HPLC	high-performance liquid chromatography
IPD	intestinal digestibility of ruminally undegraded feed crude protein
K_p	ruminal passage rate
ME	metabolizable energy
NDF	neutral detergent fiber

NEL	net energy for lactation
NPN	non-protein N
NUE	N use efficiency, i.e., proportion of milk N of total N intake
OM	organic matter
PADF	acid detergent fiber estimated from the residue after boiling in acid detergent solution according to Licitra et al. [10]
PDO	protected designation of origin
RUP	ruminally undegraded feed crude protein
RUP_{CHE}	ruminally undegraded feed crude protein estimated from chemical crude protein fractionation
RUP_{ENZ}	ruminally undegraded feed crude protein estimated from in vitro protease incubation
SEM	standard error of the mean
SI	silage
TP	true protein
uCP	utilizable crude protein at the duodenum
UNUN	urinary non-urea N
UUN	urinary urea N
VFA	volatile fatty acids

References

1. Wilkinson, J.M.; Lee, M.R.F. Review: Use of human-edible animal feeds by ruminant livestock. *Animal* **2018**, *12*, 1735–1743. [CrossRef] [PubMed]
2. Südekum, K.-H.; Krizsan, S.J.; Gerlach, K. Forage quality evaluation—Current trends and future prospects. In *The Multiple Roles of Grassland in the European Bioeconomy, Proceedings of the 26th General Meeting of the European Grassland Federation, Trondheim, Norway, 4–8 September 2016*; Höglind, M., Bakken, A.K., Hovstad, K.A., Kallioniemi, E., Riley, H., Steinshamn, H., Østrem, L., Eds.; Wageningen Academic Publishers: Wageningen, The Netherlands, 2016; pp. 151–158.
3. Anonymous. Specifications of Gruyère AOP. 2016. Available online: https://gruyere.com/content/ressources/cahier-des-charges-definitif-du-6-juillet-2001-anglais-aop-2014-v2016.pdf (accessed on 11 April 2019).
4. Bertoni, G.; Calamari, L.; Maianti, M.G. Producing specific milks for speciality cheeses. *Proc. Nutr. Soc.* **2001**, *60*, 231–246. [CrossRef] [PubMed]
5. Wyss, U.; Arrigo, Y.; Meisser, M.; Nydegger, F.; Boéchat, S.; Boessinger, M. Les facteurs de réussite de foin séché en grange à partir de l'expérience suisse. *Fourrages* **2011**, *205*, 3–10.
6. McDonald, P.; Edwards, R.A. The influence of conservation methods on digestion and utilization of forages by ruminants. *Proc. Nutr. Soc.* **1976**, *35*, 201–211. [CrossRef] [PubMed]
7. Givens, D.I.; Rulquin, H. Utilisation by ruminants of nitrogen compounds in silage-based diets. *Anim. Feed Sci. Technol.* **2004**, *114*, 1–18. [CrossRef]
8. Edmunds, B.; Südekum, K.-H.; Spiekers, H.; Schwarz, F.J. Estimating ruminal crude protein degradation of forages using in situ and in vitro techniques. *Anim. Feed Sci. Technol.* **2012**, *175*, 95–105. [CrossRef]
9. Sniffen, C.J.; O'Connor, J.D.; Van Soest, P.J.; Fox, D.G.; Russell, J.B. A net carbohydrate and protein system for evaluating cattle diets: II. Carbohydrate and protein availability. *J. Anim. Sci.* **1992**, *70*, 3562–3577. [CrossRef] [PubMed]
10. Licitra, G.; Hernandez, T.M.; Van Soest, P.J. Standardization of procedures for nitrogen fractionation of ruminant feeds. *Anim. Feed Sci. Technol.* **1996**, *57*, 347–358. [CrossRef]
11. Kirchhof, S. Kinetik des ruminalen in situ-Nährstoffabbaus von Grünlandaufwüchsen des Alpenraumes unterschiedlicher Vegetationsstadien sowie von Maissilagen und Heu—ein Beitrag zur Weiterentwicklung der Rationsgestaltung für Milchkühe. Ph.D. Thesis, Christian-Albrechts-Universität, Kiel, Germany, 2007.
12. Wilkins, R.J.; Wilkinson, J.M. Major contributions in 45 years of International Silage Conferences. In Proceedings of the XVII International Silage Conference, Piracicaba, Brazil, 1–3 July 2015; Daniel, J.L.P., Morais, G., Junges, D., Nussio, L.G., Eds.; ESALQ: Piracicaba, Brazil, 2015; pp. 26–50.
13. Grosse Brinkhaus, A.; Bee, G.; Silacci, P.; Kreuzer, M.; Dohme-Meier, F. Effect of exchanging *Onobrychis viciifolia* and *Lotus corniculatus* for *Medicago sativa* on ruminal fermentation and nitrogen turnover in dairy cows. *J. Dairy Sci.* **2016**, *99*, 4384–4397. [CrossRef] [PubMed]

14. Verband Deutscher Landwirtschaftlicher Untersuchungs-und Forschungsanstalten. *Handbuch der Landwirtschaftlichen Versuchs- und Untersuchungsmethodik (VDLUFA-Methodenbuch), Bd. III. Die Chemische Untersuchung von Futtermitteln*; VDLUFA-Verlag: Darmstadt, Germany, 2012.

15. Hall, M.B.; Hoover, W.H.; Jennings, J.P.; Miller Webster, T.K. A method for partitioning neutral detergent-soluble carbohydrates. *J. Sci. Food Agric.* **1999**, *79*, 2079–2086. [CrossRef]

16. Metzler-Zebeli, B.U.; Hooda, S.; Pieper, R.; Zijlstra, R.T.; van Kessel, A.G.; Mosenthin, R.; Gänzle, M.G. Nonstarch polysaccharides modulate bacterial microbiota, pathways for butyrate production, and abundance of pathogenic *Escherichia coli* in the pig gastrointestinal tract. *Appl. Environ. Microbiol.* **2010**, *76*, 3692–3701. [CrossRef] [PubMed]

17. Muyzer, G.; de Waal, E.C.; Uitterlinden, A.G. Profiling of complex microbial populations by denaturing gradient gel electrophoresis analysis of polymerase chain reaction-amplified genes coding for 16S rRNA. *Appl. Environ. Microbiol.* **1993**, *59*, 695–700. [PubMed]

18. Pfaffl, M.W. A new mathematical model for relative quantification in real-time RT-PCR. *Nucleic Acids Res.* **2001**, *29*, e45. [CrossRef] [PubMed]

19. Licitra, G.; Lauria, F.; Carpino, S.; Schadt, I.; Sniffen, C.J.; Van Soest, P.J. Improvement of the *Streptomyces griseus* method for degradable protein in ruminant feed. *Anim. Feed Sci. Technol.* **1998**, *72*, 1–10. [CrossRef]

20. Licitra, G.; Van Soest, P.J.; Schadt, I.; Carpino, S.; Sniffen, C.J. Influence of the concentration of the protease from *Streptomyces griseus* relative to ruminal protein degradability. *Anim. Feed Sci. Technol.* **1999**, *77*, 99–113. [CrossRef]

21. Böttger, C.; Südekum, K.-H. European distillers dried grains with solubles (DDGS): Chemical composition and in vitro evaluation of feeding value for ruminants. *Anim. Feed Sci. Technol.* **2017**, *224*, 66–77. [CrossRef]

22. Irshaid, R. Estimating Intestinal Digestibility of Feedstuffs for Ruminants Using Three-Step In Situ-In Vitro and In Vitro Procedures. Ph.D. Thesis, Christian-Albrechts-Universität, Kiel, Germany, 2007.

23. Menke, K.H.; Steingass, H. Estimation of the energetic feed value obtained from chemical analysis and in vitro gas production using rumen fluid. *Anim. Res. Dev.* **1988**, *28*, 7–55.

24. Steingaß, H.; Südekum, K.-H. Proteinbewertung beim Wiederkäuer—Grundlagen, analytische Entwicklungen und Perspektiven. *Übers. Tierernährg.* **2013**, *41*, 51–73.

25. Edmunds, B.; Südekum, K.-H.; Spiekers, H.; Schuster, M.; Schwarz, F.J. Estimating utilisable crude protein at the duodenum, a precursor to metabolisable protein for ruminants, from forages using a modified gas test. *Anim. Feed Sci. Technol.* **2012**, *175*, 106–113. [CrossRef]

26. Agroscope. Fütterungsempfehlungen für Wiederkäuer (Grünes Buch). 2016. Available online: https://www.agroscope.admin.ch/agroscope/de/home/services/dienste/futtermittel/fuetterungsempfehlungen-wiederkaeuer.html (accessed on 11 April 2019).

27. Gesellschaft für Ernährungsphysiologie. *Empfehlungen zur Energie-und Nährstoffversorgung der Milchkühe und Aufzuchtrinder*; DLG-Verlag: Frankfurt am Main, Germany, 2001.

28. Carter, W.R.B. A review of nutrient losses and efficiency of conserving herbage as silage, barn-dried hay and field-cured hay. *Grass Forage Sci.* **1960**, *15*, 220–230. [CrossRef]

29. McGechan, M.B. A review of losses arising during conservation of grass forage: Part 1, field losses. *J. Agric. Eng. Res.* **1989**, *44*, 1–21. [CrossRef]

30. Deutsche Landwirtschafts-Gesellschaft. Grobfutterbewertung-Teil B—DLG-Schlüssel zur Beurteilung der Gärqualität von Grünfuttersilagen auf Basis der chemischen Untersuchung. DLG-Information 2/2006. 2006. Available online: https://www.dlg.org/fileadmin/downloads/fachinfos/futtermittel/grobfutterbewertung_B.pdf (accessed on 11 April 2019).

31. Gerlach, K.; Roß, F.; Weiß, K.; Büscher, W.; Südekum, K.-H. Aerobic exposure of grass silages and its impact on dry matter intake and preference by goats. *Small Rumin. Res.* **2014**, *117*, 131–141. [CrossRef]

32. Campling, R.C. The intake of hay and silage by cows. *Grass Forage Sci.* **1966**, *21*, 41–48. [CrossRef]

33. Thiago, L.R.L.; Gill, M.; Dhanoa, M.S. Studies of method of conserving grass herbage and frequency of feeding in cattle: 1. Voluntary feed intake, digestion and rate of passage. *Br. J. Nutr.* **1992**, *67*, 305–318. [CrossRef] [PubMed]

34. Huhtanen, P.; Rinne, M.; Nousiainen, J. Evaluation of the factors affecting silage intake of dairy cows: A revision of the relative silage dry-matter intake index. *Animal* **2007**, *1*, 758–770. [CrossRef] [PubMed]

35. Demarquilly, C.; Jarrige, R. The effect of method of forage conservation on digestibility and voluntary intake. In Proceedings of the XI International Grassland Congress, Surfers Paradise, Australia, 13–23 April 1970; Norman, M.J.T., Ed.; University of Queensland Press: St. Lucia, Australia, 1970; pp. 733–737.

36. Verbič, J.; Ørskov, E.R.; Žgajnar, J.; Chen, X.B.; Žnidaršič-Pongrac, V. The effect of method of forage preservation on the protein degradability and microbial protein synthesis in the rumen. *Anim. Feed Sci. Technol.* **1999**, *82*, 195–212. [CrossRef]

37. Friggens, N.C.; Oldham, J.D.; Dewhurst, R.J.; Horgan, G. Proportions of volatile fatty acids in relation to the chemical composition of feeds based on grass silage. *J. Dairy Sci.* **1998**, *81*, 1331–1344. [CrossRef]

38. Jaakkola, S.; Huhtanen, P. The effects of forage preservation method and proportion of concentrate on nitrogen digestion and rumen fermentation in cattle. *Grass Forage Sci.* **1993**, *48*, 146–154. [CrossRef]

39. Shingfield, K.J.; Jaakkola, S.; Huhtanen, P. Effect of forage conservation method, concentrate level and propylene glycol on diet digestibility, rumen fermentation, blood metabolite concentrations and nutrient utilisation of dairy cows. *Anim. Feed Sci. Technol.* **2002**, *97*, 1–21. [CrossRef]

40. Stevenson, D.M.; Weimer, P.J. Dominance of *Prevotella* and low abundance of classical ruminal bacterial species in the bovine rumen revealed by relative quantification real-time PCR. *Appl. Microbiol. Biotechnol.* **2007**, *75*, 165–174. [CrossRef]

41. Hartinger, T.; Gresner, N.; Südekum, K.-H. Does intra-ruminal nitrogen recycling waste valuable resources? A review of major players and their manipulation. *J. Anim. Sci. Biotechnol.* **2018**, *9*, 33. [CrossRef] [PubMed]

42. Castillo, A.R.; Kebreab, E.; Beever, D.E.; Barbi, J.H.; Sutton, J.D.; Kirby, H.C.; France, J. The effect of protein supplementation on nitrogen utilization in lactating dairy cows fed grass silage diets. *J. Anim. Sci.* **2001**, *79*, 247–253. [CrossRef] [PubMed]

43. Mulligan, F.J.; Dillon, P.; Callan, J.J.; Rath, M.; O'Mara, F.P. Supplementary concentrate type affects nitrogen excretion of grazing dairy cows. *J. Dairy Sci.* **2004**, *87*, 3451–3460. [CrossRef]

44. Olmos Colmenero, J.J.; Broderick, G.A. Effect of dietary crude protein concentration on milk production and nitrogen utilization in lactating dairy cows. *J. Dairy Sci.* **2006**, *89*, 1704–1712. [CrossRef]

45. Pfeffer, E.; Schuba, J.; Südekum, K.-H. Nitrogen supply in cattle coupled with appropriate supply of utilisable crude protein at the duodenum, a precursor to metabolisable protein. *Arch. Anim. Nutr.* **2016**, *70*, 293–306. [CrossRef]

46. Spek, J.W.; Dijkstra, J.; van Duinkerken, G.; Bannink, A. A review of factors influencing milk urea concentration and its relationship with urinary urea excretion in lactating dairy cattle. *J. Agric. Sci.* **2013**, *151*, 407–423. [CrossRef]

47. Castillo, A.R.; Kebreab, E.; Beever, D.E.; France, J. A review of efficiency of nitrogen utilisation in lactating dairy cows and its relationship with environmental pollution. *J. Anim. Feed Sci.* **2000**, *9*, 1–32. [CrossRef]

48. Volden, H.; Mydland, L.T.; Olaisen, V. Apparent ruminal degradation and rumen escape of soluble nitrogen fractions in grass and grass silage administered intraruminally to lactating dairy cows. *J. Anim. Sci.* **2002**, *80*, 2704–2716. [CrossRef]

49. Totty, V.K.; Greenwood, S.L.; Bryant, R.H.; Edwards, G.R. Nitrogen partitioning and milk production of dairy cows grazing simple and diverse pastures. *J. Dairy Sci.* **2013**, *96*, 141–149. [CrossRef]

50. Huhtanen, P.; Hristov, A.N. A meta-analysis of the effects of dietary protein concentration and degradability on milk protein yield and milk N efficiency in dairy cows. *J. Dairy Sci.* **2009**, *92*, 3222–3232. [CrossRef]

51. Scherer, R.; Gerlach, K.; Taubert, J.; Adolph, S.; Weiß, K.; Südekum, K.-H. Effect of forage species and ensiling conditions on silage composition and quality and the feed choice behaviour of goats. *Grass Forage Sci.* **2019**. In press. [CrossRef]

52. Edmunds, B.; Spiekers, H.; Südekum, K.-H.; Nussbaum, H.; Schwarz, F.J.; Bennett, R. Effect of extent and rate of wilting on nitrogen components of grass silage. *Grass Forage Sci.* **2014**, *69*, 140–152. [CrossRef]

53. Johansen, M.; Hellwing, A.L.F.; Lund, P.; Weisbjerg, M.R. Metabolisable protein supply to lactating dairy cows increased with increasing dry matter concentration in grass-clover silage. *Anim. Feed Sci. Technol.* **2017**, *227*, 95–106. [CrossRef]

54. Hoekstra, N.J.; Schulte, R.P.O.; Struik, P.C.; Lantinga, E.A. Pathways to improving the N efficiency of grazing bovines. *Eur. J. Agron.* **2007**, *26*, 363–374. [CrossRef]

55. Frydrych, Z. Intestinal digestibility of rumen undegraded protein of various feeds as estimated by the mobile bag technique. *Anim. Feed Sci. Technol.* **1992**, *37*, 161–172. [CrossRef]

56. Cone, J.W.; van Gelder, A.H.; Mathijssen-Kamman, A.A.; Hindle, V.A. Post-ruminal digestibility of crude protein from grass and grass silages in cows. *Anim. Feed Sci. Technol.* **2006**, *128*, 42–52. [CrossRef]

57. Edmunds, B.; Südekum, K.-H.; Bennett, R.; Schröder, A.; Spiekers, H.; Schwarz, F.J. The amino acid composition of rumen-undegradable protein: A comparison between forages. *J. Dairy Sci.* **2013**, *96*, 4568–4577. [CrossRef]

58. Calsamiglia, S.; Ferret, A.; Reynolds, C.K.; Kristensen, N.B.; van Vuuren, A.M. Strategies for optimizing nitrogen use by ruminants. *Animal* **2010**, *4*, 1184–1196. [CrossRef]

59. Huhtanen, P.; Nousiainen, J.I.; Rinne, M.; Kytölä, K.; Khalili, H. Utilization and partition of dietary nitrogen in dairy cows fed grass silage-based diets. *J. Dairy Sci.* **2008**, *91*, 3589–3599. [CrossRef]

60. Nadeau, E.; Englund, J.-E.; Gustafsson, A.H. Nitrogen efficiency of dairy cows as affected by diet and milk yield. *Livest. Sci.* **2007**, *111*, 45–56. [CrossRef]

61. Tas, B.M.; Taweel, H.Z.; Smit, H.J.; Elgersma, A.; Dijkstra, J.; Tamminga, S. Effects of perennial ryegrass cultivars on milk yield and nitrogen utilization in grazing dairy cows. *J. Dairy Sci.* **2006**, *89*, 3494–3500. [CrossRef]

62. Swick, R.W.; Benevenga, N.J. Labile protein reserves and protein turnover. *J. Dairy Sci.* **1977**, *60*, 505–515. [CrossRef]

63. Komaragiri, M.V.S.; Erdman, R.A. Factors affecting body tissue mobilization in early lactation dairy cows. 1. Effect of dietary protein on mobilization of body fat and protein. *J. Dairy Sci.* **1997**, *80*, 929–937. [CrossRef]

64. Tamminga, S.; Luteijn, P.A.; Meijer, R.G.M. Changes in composition and energy content of liveweight loss in dairy cows with time after parturition. *Livest. Prod. Sci.* **1997**, *52*, 31–38. [CrossRef]

65. Sinclair, L.A.; Garnsworthy, P.C.; Newbold, J.R.; Buttery, P.J. Effect of synchronizing the rate of dietary energy and nitrogen release on rumen fermentation and microbial protein synthesis in sheep. *J. Agric. Sci.* **1993**, *120*, 251–263. [CrossRef]

66. Rius, A.G.; McGilliard, M.L.; Umberger, C.A.; Hanigan, M.D. Interactions of energy and predicted metabolizable protein in determining nitrogen efficiency in the lactating dairy cow. *J. Dairy Sci.* **2010**, *93*, 2034–2043. [CrossRef] [PubMed]

67. Tamminga, S. A review on environmental impacts of nutritional strategies in ruminants. *J. Anim. Sci.* **1996**, *74*, 3112–3124. [CrossRef]

68. Zbinden, R.S.; Falk, M.; Münger, A.; Dohme-Meier, F.; van Dorland, H.A.; Bruckmaier, R.M.; Gross, J.J. Metabolic load in dairy cows kept in herbage-based feeding systems and suitability of potential markers for compromised well-being. *J. Anim. Physiol. Anim. Nutr.* **2017**, *101*, 767–778. [CrossRef]

69. Dewhurst, R.J.; Mitton, A.M.; Offer, N.W.; Thomas, C. Effects of the composition of grass silages on milk production and nitrogen utilization by dairy cows. *Anim. Sci.* **1996**, *62*, 25–34. [CrossRef]

70. Kolver, E.S.; Muller, L.D. Performance and nutrient intake of high producing Holstein cows consuming pasture or a total mixed ration. *J. Dairy Sci.* **1998**, *81*, 1403–1411. [CrossRef]

agriculture

MDPI

Communication

The Effects of Forage Policy on Feed Costs in Korea

Jae Bong Chang

Department of Food Marketing and Safety, Konkuk University, Seoul 05029, Korea; jbchang@konkuk.ac.kr;
Tel.: +82-2-2049-6010

Received: 30 April 2018; Accepted: 27 May 2018; Published: 29 May 2018

Abstract: Feeding operations are substantial on livestock farms, besides being potentially expensive. Feeding efficiency has been considered a major influence on profits in the livestock industry. Indeed, feed costs are shown to be the largest single item of production cost in Korea. To promote production and use of domestic forage, the Korean government has enforced the forage base expansion program that strengthens the competitiveness of the livestock industry by reducing the production cost. The forage base expansion program includes three main policies: subsidized forage production, support for processing and distribution, and expanding land for forage production. This paper investigates the influence of the government's policies often conjectured to have pronounced effects on forage production. To evaluate the forage policies, this paper uses a path-analysis approach linking government spending on forage base expansion programs and feed costs. Results indicate that the Korean government's spending on supporting domestic forage production results in a decrease in the ratio of forage expenses to total feed cost.

Keywords: feed costs; forage production; path analysis; policy

1. Introduction

Many countries are highly dependent on feed imports and, in general, feed is the largest part of production costs. For example, Korea imports 75% of its compound feed and 96.4% of its feed crops, which has become a matter of concern among Korean livestock industry participants and the Korean government [1]. Feed is the most significant cost in livestock production, often representing more than half of the production costs. Indeed, the portion of feed cost for Hanwoo (Korean beef cattle) and dairy cattle are 38% and 58% of the total cost to produce beef, respectively [2]. Thus, the price of international crops and the surge in feed prices related to oil prices directly affect domestic livestock farms.

The livestock industry contributes more than 40% of the total value of agricultural output in Korea. From the beginning of 1990 to 2016, consumption of meat in Korea increased by 1.7 million tons. During the same period, per-capita meat consumption increased from 19.9 kg to 49.5 kg [3]. Rising domestic consumption has been an important factor in stimulating import demand, resulting in the self-sufficiency of meat production decreasing from 90.0% to 68.0%. Furthermore, the beef self-sufficiency rate is much lower than total meat, and it recorded 41% in 2016.

There has been a trend towards fewer livestock farms with large numbers of animals to achieve economies of scale. For instance, the average number of Hanwoo and dairy cattle per farm in 2017 was 31.6, which is about 10 times that of 1990. As of 2017, Korea had a total of 3.0 million beef cattle and 409,000 dairy cows.

Making a small change in the quantity of the type of feed has a greater impact on profitability than any other cost due to its large impact on production costs. To operate stable animal husbandry and production management, producing high-quality forage can reduce feed costs and create an import substitution effect. As a result, the Korean Ministry of Agriculture, Food and Rural Affairs (MAFRA) has implemented the forage base expansion program in 1998 to expand the production base for domestic forage and to utilize resources.

The forage base expansion program includes three main policies: subsidized forage production, support for processing and distribution of forage, and expanding land for forage production. Firstly, the vast majority of spending on the program, 65 percent, is allocated to forage production subsidies such as the silage production, forage seed, and equipment and machine. For instance, the Korean government supports the cost of silage production by 60,000 KRW/ton (US$50.72/ton based on the exchange rate of 1182.91 KRW/US$ on January 2017). Secondly, the regional hub and distribution center or total mixed ration (TMR) suppliers are supported to allow efficient supply of raw materials. In 2017, the Korean government spent a total of 45 billion KRW on supporting the processing and distribution facilities for forage production. Thirdly, the government creates and expands the specialized zone for forage production by subsidizing raw materials such as silage, seed, equipment, machinery and compost.

Although there have been many arguments on the merits of supporting domestic forage production, relatively little empirical work has investigated the effects of this program. In general, policies and regulations can affect domestic forage production directly or indirectly. Since one of the primary issues associated with forage policy is identifying how this public program affects the livestock industry in Korea, it is important to determine how the forage production expansion policy affects domestic forage production and how this domestically produced forage impacts feed costs. Thus, the purpose of this article is to examine the effects of the aforementioned policy to support domestic forage production. This study uses a path-analysis model linking government spending on the forage base expansion program to feed costs, and disaggregates the effects to different stages. The results of this study might provide insight into the process of forage policy influencing feed costs in Korea and initial information to support better understanding for adaptive strategies by other countries.

2. Materials and Methods

2.1. Path Analysis

To accomplish the research objectives, the hypothesized relationship between government spending on expanding domestic forage production and feed costs was examined using a path-analysis approach. Path analysis refers to a variety of statistical techniques that aim to represent the directed dependencies among a set of observable variables and to test the causal interactions among variables [4]. The advantage of a path analysis over regression is that it concurrently performs multiple regression analyses while it produces an overall assessment of the fit of the model, usually based on a χ^2 statistic [5]. This type of so-called sequential multiple regression model is widely used in policy analysis [6,7] and has been discussed in detail by authors such as Dye and Pollack [8], Edwards and Lambert [9], and Pedhazur [10].

This study builds on the simple path model, focuses on forage-related policy, and links government spending with feed costs. More specifically, the share of forage feed cost in total feed costs is a dependent variable and the exogenous independent variable is expenditure on forage production. The assumption is that there are only three endogenous independent variables, the domestic forage production, forage imports, and head of beef cattle. The primary focus is on how the policy related to the use of government spending might affect the livestock industry associated with feed costs, rather than on methodological or theoretical issues. Thus, for this study, the path model was estimated using three main sets of maximum likelihood estimation equations in the CALIS SAS procedure, as shown in Figure 1. The CALIS procedure uses a variety of modeling languages to fit structural equation models. The first set explored the relationship between government spending (*Budget*) in year t as an independent variable, and domestic forage production (*Domestic*) in year t and forage imports (*Imports*) in year t as dependent variables, and these are given by the Equation (1).

$$Domestic_t = \alpha_0 Budget_t + e1_t, \text{ and } Import_t = \alpha_1 Budget_t + e2_t, \tag{1}$$

where e_t are the error terms. The second set, in which the head of beef cattle (*Cattle*) in year t was the dependent variable, included forage imports and domestic forage production as independent variables, and is expressed as the Equation (2).

$$Cattle_t = \beta_0 Domestic_t + \beta_1 Import_t + e3_t \qquad (2)$$

In the last step, the head of beef cattle is the basic factor influencing the share of forage feed costs in the total feed costs (*Share_Fcost*). Besides, the path model defined the share of forage feed cost as the dependent variable, and public spending as well as the domestic forage production, forage imports, and the head of beef cattle as independent variables. The path model for the share of forage feed costs is shown as the Equation (3).

$$Share\_Fcost_t = \gamma_0 Budget_t + \gamma_1 Domestic_t + \gamma_2 Import_t + \gamma_3 Cattle_t + e4_t \qquad (3)$$

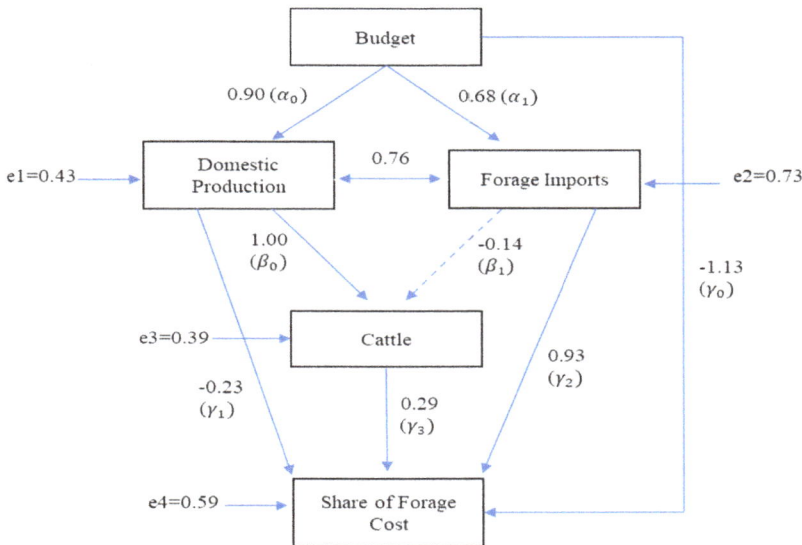

Figure 1. Path-analysis diagram for how feed cost is affected by government spending on domestic forage production in Korea, from 1998 to 2016. All the lines in the diagram represent a specific linear model. Solid lines indicate good evidence for an effect (95% or higher significance), and dotted lines indicate no clear relationship.

2.2. Data

Table 1 shows the descriptions of all variables and the descriptive statistics for path analysis. Data on the budget for expanding forage production by the government are inferred from the MAFRA spending on the forage base expansion program over the years 1998 to 2016. Over time, total government costs have trended upward and recorded an annual growth rate of 9.9% during this period. Total government costs of the forage base expansion program in 2016 were 118,997 million KRW. Driven in part by the increase in government spending on the forage base expansion program, domestic forage production increased annually, by 1.5%, from 1998 to 2016. While domestic forage production increased, this change remained below the eighteen-year historical average increase of 13.4% in forage imports. The annual growth rates of head of beef cattle, feed costs and forage feed costs were 1.2%, 6.6% and 7.3% respectively.

Table 1. Descriptive statistics for forage and beef cattle industry in Korea: 1998–2016.

Variable	Descriptions	Mean	Standard Deviation	Min.	Max.	Annual Growth Rate
Budget	Annual government spending on the forage base expansion program in 1 million KRW	68,348.1	50,444.8	15,600	157,707	9.9%
Domestic	Domestic forage production in 1000 tons	3859.4	590.9	2793	4672	1.5%
Import	Forage import in 1000 tons	795.1	259.4	172	1120	13.4%
Cattle	Number of beef cattle heads in 1000s	2262.8	578.0	1406	3059	1.2%
Cost	Annual feed cost per head in KRW	1,832,697.8	862,191.6	803,038	2,982,290	6.6%
Fcost	Annual cost for forage per head in KRW	282,637.0	123,665.4	114,257	452,739	7.3%

Source, Ministry of Agriculture, Food and Rural Affairs (2017), Statistics Korea (2017).

3. Results

The results from the estimated path model are reported in Table 2 (and also Figure 1). These coefficient values can also be observed in Figure 1. Estimates associated with variables can be interpreted as the relative strength and a sign of the causal effect to outcome variables. With only one exception, all parameter estimates were statistically significant and carried the expected signs. In addition, the goodness-of-fit index of 0.91 indicates a reasonably good fit of the path model, and a chi-squared value of 4.8313 with two degrees of freedom yields a *p*-value of greater than 0.05. Thus, the path model is not rejected.

The government spending ($\alpha_0 = 0.903$) had a positive direct effect on expanding domestic forage production. The results also present that domestic forage production ($\beta_0 = 0.997$) had a positive effect on the number of beef cattle herds. The explanatory variables accounted for 85% of the variance in explaining health behaviors. In addition, in the final model, government spending ($\gamma_0 = -1.128$) and domestic forage production ($\gamma_1 = -0.225$) had a negative effect, while beef cattle herds ($\gamma_3 = 0.291$) and forage imports ($\gamma_4 = 0.930$) had a positive effect on the share of the forage cost to the total feed cost. The model variables accounted for 65% of the variance of the share of forage cost in the total feed cost.

Table 2. Path model estimates.

Outcome	R^2	Variable	Coefficients	Standard Error	*p*-Value
Share_Fcost	0.65	Budget	−1.128	0.027	<0.001
		Cattle	0.291	0.038	<0.001
		Domestic	−0.225	0.022	<0.001
		Import	0.930	0.017	<0.001
Cattle	0.85	Domestic	0.997	0.064	<0.001
		Import	−0.135	0.117	0.248
Domestic	0.82	Budget	0.903	0.019	<0.001
Import	0.47	Budget	0.682	0.092	<0.001

Government expenditure toward expanding forage production had a significant impact on domestic forage production, which in turn decreased the share of forage cost in the total feed cost. The Pearson correlation (0.76) shows that the domestic forage production correlates to forage imports positively. It is worth emphasizing the difference in the estimated effects of domestic forage production and imports to the forage total feed–cost ratio of beef cattle, which presumably reflects underlying differences in the feed costs implied by the two different forage feeds. It also implies that the two forage feeds are substitutes.

The direct and indirect effects of government intervention are summarized in Table 3. The results indicated that government expenditure had the largest direct effect on the share of forage cost in total feed costs. Apart from the direct effect on the forage feed costs ratio, government expenditure had a positive indirect effect on the share of forage feed costs through their impacts on forage imports and the number of beef cattle. Domestic forage production was found to lower the share of forage cost in total feed costs.

Table 3. Estimated direct and indirect effects of government spending and domestic forage production on cost share of forage in the path analysis model.

Factors of the Share of Forage Cost	Direct Effect	Indirect Effect	Total Effect
Budget	−1.128 **	0.667 **	−0.462 **
	(0.082)	(0.204)	(0.123)
Domestic	−0.225 **	0.291 **	0.065 *
	(0.022)	(0.052)	(0.036)

Significance level: ** p-value < 0.05. * p-value < 0.10; numbers in parentheses ( ) are standard error.

4. Discussion

Livestock production costs are gaining attention in the public and private sectors in Korea. In Korean livestock farming, the self-sufficient feed ratio has declined each year and the high level of feed imports has increased production costs. There are several reasons for the decreased or stagnant self-sufficiency ratio in Korea. Because the primary arable land use is for crop, mainly rice, production, there is high competition with crops and livestock. This must necessarily lead to replacement of one by the other, or specialization. Furthermore, livestock farmers in Korea tend to consider the quality of imported forage better than locally produced forage. Environmental factors such as temperatures, precipitation or location are the fundamental factors influencing the forage quality. However, four seasons in Korea do not always allow forage to achieve the equal levels of quality. Another reason might be related to the increase in the scale of livestock and rice production. Where the average farm size has increased, mechanized and specialized farming systems have become dominant, and these depend on concentrated and imported forage feed for livestock farming.

As a result, several production cost-reduction initiatives have emerged in recent years and the Korean MAFRA has already been moving toward implementing some of these policies, particularly policy solutions toward reducing feed costs. For example, the forage base expansion program that encourages domestic forage production of import-substituting forage has been implemented since 1998. Spending on supporting domestic forage production varies from year to year, but it has hovered around 120 billion KRW (US$ 100 million) in recent years (note that average exchange rate for Korean won (KRW) to USD in 2016 was 1 USD = 1160.50 KRW (Bank of Korea)). Despite the fact that, on average, public spending for domestic forage production has been increasing, the impact of this government intervention remains unclear. While the primary aim of the forage production policy is to boost domestic products, it also induces reducing feed cost.

Applying the path-analysis model, the effects of government spending on the domestic forage base expansion program and the share of forage feed costs were examined. The public spending on domestic forage production is directly (positively) related to the expansion of domestic forage production, which is consistent with findings from a previous study [11], and in turn, aims to decrease the share of forage feed costs. Ahn and Han [11] indicated that Korea's forage self-sufficiency rate would remain about 56 percent with no government support. Despite this effect, there is not enough of a substitution effect between imported and domestically produced forage. As total spending for the forage expansion program by the government has trended upward over time, domestically produced forage product has increased but has averaged only 1.5 percent per year for the past eighteen years.

However, on average, the annual growth rate of imports has been as high as 13.4%, 11.9 percentage points higher than the 1.5% annual domestic production growth rate over the corresponding period. Thus, since 2010, Korea's forage self-sufficiency has remained fairly stable each year at approximately 80 percent.

According to a MAFRA report [12], on average, prices of domestic forage were about 14.3% lower on a dry-matter basis than the price of import forage, which is mainly caused by shipping and handling costs. In addition, rice has been the most valuable crop and constitutes a major source of farm income in Korea. However, with the change of Korean diet and eating habits, per-capita annual rice consumption showed a continuously decreasing trend. To address this issue, the Korean government began a program which aims to change cropping systems for paddy fields with food or forage crops alternative to rice.

Korean livestock farmers will be more unprosperous when Korea further opens its agricultural markets to competitors all over the world by having freer trade. Ensuring a flow of efficiency-enhancing animal husbandry is a mechanism to improve farmer and livestock industry wellbeing. Economic and environmental goals are important, and many can be reached through government policies that are encouraging farms to convert from food to feed, providing subsidies for both the growing forage as well as to their purchases by beef and dairy cattle industries and in more local ways.

Conflicts of Interest: The author declares no conflict of interest.

References

1. Sung, M.H.; Yoon, J. *Status of Feedstuffs Imports and Calculation of Import Price Index*; Korea Rural Economic Institute: Seoul, Korea, 2013.
2. Statistics Korea. *Livestock Production Cost Survey Report 2016*; Statistics Korea: Seoul, Korea, 2017.
3. Korean Ministry of Agriculture, Food and Rural Affairs. *Statistics of Agriculture, Livestock and Food 2017*; Ministry of Agriculture, Food and Rural Affairs: Sejong, Korea, 2017.
4. Smith, A.F.; Brown, H.J.; Valone, J.T. Path analysis: A critical evaluation using long-term experimental data. *Am. Nat.* **1997**, *149*, 29–42. [CrossRef]
5. Singh, J.; Wilkes, R.E. When consumers complain: A path analysis of the key antecedents of consumer complaint response estimates. *J. Acad. Mark. Sci.* **1996**, *24*, 350–365. [CrossRef]
6. Defourny, J.; Thorbecke, E. Structural path analysis and multiplier decomposition within a social accounting matrix framework. *Econ. J.* **1984**, *94*, 111–136. [CrossRef]
7. Erdag, C. Accountability policies at schools: A study of path analysis. *Educ. Sci. Theor. Pract.* **2017**, *17*, 1405–1432. [CrossRef]
8. Dye, T.R.; Pollack, N.F. Path analytic models in policy research. *Policy Stud. J.* **1973**, *2*, 123–130. [CrossRef]
9. Edwards, J.R.; Lambert, L.S. Methods for integrating moderation and mediation: A general analytical framework using moderate path analysis. *Psychol. Methods* **2007**, *12*, 1–22. [CrossRef] [PubMed]
10. Pedhazur, E.J. *Multiple Regression in Behavioral Research*, 3rd ed.; Cengage Learning: New York, NY, USA, 1997; p. 1072, ISBN 9780030728310.
11. Ahn, B.I.; Han, S.H. Analysis on the effects of government's support for forage production. *Korean J. Agric. Econ.* **2016**, *57*, 55–78.
12. Korean Ministry of Agriculture, Food and Rural Affairs. Demonstration of Harvesting Grass Feed, Promoting Production and Utilization. Available online: http://www.korea.kr/common/download.do?fileId=184244049&tblKey=GMN (accessed on 11 May 2016).

MDPI

St. Alban-Anlage 66

4052 Basel

Switzerland

Tel. +41 61 683 77 34

Fax +41 61 302 89 18

www.mdpi.com

Agriculture Editorial Office

E-mail: agriculture@mdpi.com

www.mdpi.com/journal/agriculture

www.ingramcontent.com/pod-product-compliance
Lightning Source LLC
Chambersburg PA
CBHW051908210326
41597CB00033B/6065